Collector's Guide to

Electric FANS

Identification and Values

John M. Witt

COLLECTOR BOOKS
A Division of Schroeder Publishing Co., Inc.

Searching For A Publisher?

Cover Information:

Top: ***R & M***	*Center:* ***Emerson***	*Bottom Left:* ***GE***	*Bottom Right:* ***Cold Spot***
Type: D10-A6-01	*Type: 5250-B Silver Swan*	*Type: unknown*	*Type: unknown*
Blades: 10"	*Blades: 10"*	*Blades: 10"*	*Blades:*
Date: 1940	*Date: 1934*	*Date: 1905*	*Date: 1930*
Grade I – $130.00	Grade I – $300.00	Grade I – $350.00	Grade I – $70.00
Grade II – $96.00	Grade II – $225.00	Grade II – $263.00	Grade II – $53.00
Grade III – $65.00	Grade III – $150.00	Grade III – $175.00	Grade III – $35.00
Grade IV – $46.00	Grade IV – $105.00	Grade IV – $123.00	Grade IV – $25.00
Grade V – $26.00	Grade V – $60.00	Grade V – $70.00	Grade V – $14.00
Grade VI – $7.00	Grade VI – $15.00	Grade VI – $18.00	Grade VI – $3.00

Cover design by Terri Stalions
Book design by Beth Ray

Additional copies of this book may be ordered from:

COLLECTOR BOOKS
P.O. Box 3009
Paducah, Kentucky 42002-3009

@$16.95. Add $2.00 for postage and handling.

Contents

Dedication

I lovingly dedicate this book to my parents and sister.

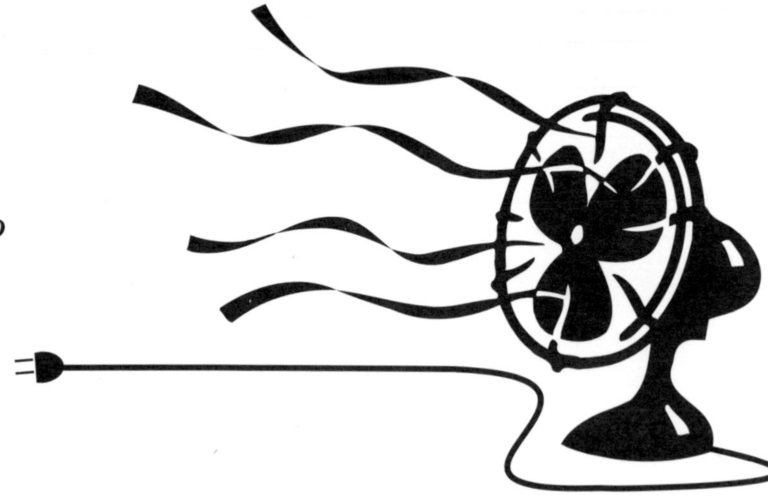

Acknowledgments

There are many people that I would like to thank in the development of this book. I would like to thank Warren Kernell, Dick Bosewell, and John Trier for allowing me to photograph their collections for this book. I would like to especially thank Jim McLaughlin for editing my book and giving me encouragement. I would also like to thank Virgil L. Babcock for his computer work on the book at a very critical time.

About the Author

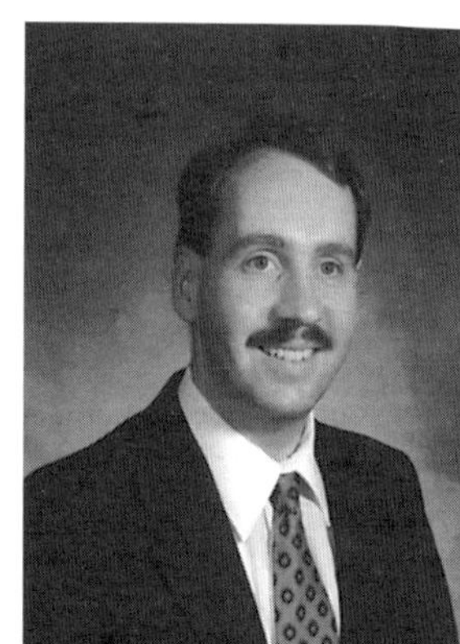

John is a native of St. Louis, Missouri, and grew up loving old electric fans. He began collecting when he was ten years old. By the age of thirteen he had overy thirty fans and had been written up in the *St. Louis Post Dispatch* as a young collector of this unusual hobby. He causally collected fans for the next fifteen years until discovering the American Fan Collectors Association. Following the revelation that there were many others interested in old electric fans he began actively collecting again. In 1991 John published his first book on electric desk fans and has written several articles on the subject for several antique journals. He is a librarian by trade and a graduate of four universities. He resides in historic St. Joseph, Missouri.

Introduction

Americans are great collectors and we collect anything and everything. Electric desk fans are no exception. Fans are different from most other collectibles in that they are complex machines with many working parts. There are many individual aspects that affect the value of a fan such as age, style, condition, and manufacturer to name just a few. Unlike glassware or pottery which is either perfect or not, fans have many variables that affect a fan's value. The most important aspect in selecting a fan may be that you personally like it and just want to have it in your collection.

As with collecting anything, the lack of sufficient knowledge can prove disastrous for the amateur. Unscrupulous individuals can assemble fans with counterfeit parts and sell them to unsuspecting collectors. These asembled fakes are close to worthless and only tend to confuse the collector. Dating fans can be tricky since many fans were manufactured for multiple years and some manufacturers borrowed parts from each other.

This book is a guide of general information concerning desk fans. It does not cover ceiling or non-mechanical fans. It is designed to establish some parameters within the field of desk fan collecting. For more information about fans and a bi-monthly newsletter about fans, write:

AFCA
P.O. Box 5473
Sarasota, FL 34277-5473

Dating Fans

In this book you will see that each fan has been dated. Each date of a particular fan will follow the size of the blades in each listing. There will be two kinds of date listings in this book. The first type will be a confirmed date supported by information found in the literature. This date will look as follows: Date: 1919. The second type of date will be an approximate date of production based on the author's own experience and review of the literature. This date will look as follows: Date: c.1919. Dates used in this book are to give the collector an approximate time of manufacture. It should be noted that many fans had multiple years of manufacture which makes exact dating difficult, especially for post World War I fans. I have tried to point out some of these differences in this book. It should be also noted that not all fans in this book are in chronological order, this is due to the way the pictures were taken by the author.

How to Use this Value Guide

Fans in this book have been carefully priced to meet the description of each "grade" category. When pricing a fan from this book you need to look over the fan carefully and then decide which grade category the fan falls into. Most fans purchased will fall into Grade III or lower. Only a very few select fans are ever found in mint or Grade I condition. Fans made before 1915 are almost never found in Grade I condition. Antique fans found in "mint" condition are very valuable and the prices listed are the author's estimations. The actual price for a mint condition fan will be the agreed upon price between the seller and buyer. In some areas fans are very plentiful and common, in other areas they are few and far between. Read each grade category carefully and then make your decision. The older the fan the more likely it will be in a grade IV or lower.

Grading and Pricing

Grade I — Mint Condition: original paint with no blemishes and in perfect original condition, brass and/or chrome parts in original polished or lacquered state with no blemishes, original cloth or rubber cord in perfect condition with no cracks or frays and having original plug, all moving parts in excellent working order, head wire in perfect condition with no cracks or frays, no cracks or dents of any kind in the castings or stamped steel part, all screws original and in their correct place, felt base original and in perfect condition, internal parts such as switch, speed coil, stator, field coils in like new perfect working condition. The fan would be like it just came from the factory.

Grade II — Professional Restoration: new paint to match original finish in perfect restored condition, brass and/or chrome parts polished and lacquered, replacement cord as close to original as possible with original type plug, all moving parts in excellent and working order, head wired original or replaced with similar cord, no cracks or dents of any kind in the castings or stamped steel parts, felt on the base should be replaced in the original color, internal parts may be replaced such as new field coils or switch all in working condition. The fan should give the appearance it just came from the factory.

Grade III — Excellent Condition: original paint with few blemishes and in clean condition, brass and/or chrome parts in clean, polished attractive condition with few noticeable blemishes, original cord intact and useable or, with replacement cord similar to the original, all moving parts in working order, head wire original in good working condition, castings and stamped steel parts in good condition with no apparent blemishes, original screws or manufactured replacement screws in excellent condition, internal parts such as switch, speed coil, stator, field coils in good, clean workable order.

Grade IV — Good Condition: original paint with some blemishes and chips, there may be some rust bleeding though the finish, brass and/or chrome parts all present in dirty and soiled condition, all moving parts should be in working order or with a few minor adjustments (like lubricating) become operable, castings and stamped steel parts should all be there and may have slight cracks or dents, most screws should be original and internal parts such as switch, speed coil, stator, field coils in working condition.

Grade V — Poor Condition: original paint dirty, scratched, and in poor condition, rust bleeding through finish, brass and/or chrome parts in dirty and dull condition, or may be missing, original cord missing or badly damaged, moving parts may be in working order, head wire frayed or missing, screws damaged and/or missing, internal parts in dirty, unrestored condition and may be in working order, parts may be missing or badly damaged.

Grade VI — Scrap fan: missing major parts, finish is rusty, dirty condition and original paint not salvageable, brass and/or chrome parts missing or badly damaged, cord and head wire missing, screws missing, moving parts not in working order, internal parts may be missing or badly damaged.

Emerson Electric

The Emerson Electric and Manufacturing Company of St. Louis, Missouri, manufactured the highest quality fans made in the United States. They manufactured fans from 1891 until the mid-1960s and are still located in St. Louis.

The company began on September 17, 1890, when Judge John Wesley Emerson and the Meston brothers, Alexander and Charles met in St. Louis to form the Emerson Electric and Manufacturing Company. A week later they filed for incorporation with the State of Missouri with $50,000 capital and a building at 904 Olive Street in St. Louis.

Emerson began marketing their first Meston Electric Alternating Motor fan late in 1891. The Meston brothers had been inventing for over 10 years and held several important patents with the United States Patent Office. Alex Meston filed for his alternating current motor on June 15, 1891, and the Meston motor was official. This new motor had other applications such as for dentists' drills, jewelers' buffers, and running sewing machines.

Emerson developed four desk fan models in the 1890s, the first being the Meston fan and the last being the PI 241. In 1897 they marketed their first ceiling fan and this line of new fans turned out to be very profitable for the company and was responsible for 50% of all fan sales by 1900.

Herbert L. Parker became president of Emerson in 1892, and was a very effective leader in the early development of Emerson fans. He patented the scalloped Parker Blade on September 12, 1899, which became a hallmark of Emerson fans.

Emerson developed two lines of fans early in the twentieth century: the top grade called Emerson and less expensive grade called Trojan. The Trojan line of fans came out in 1904, the year of the St. Louis World's Fair and were manufactured until about 1912. Trojan fans did not have a single bearing motor or Parker blades, and these fans were manufactured for the lower-priced market.

In 1909 Emerson introduced their first 8" small fan and the following year introduced the first six-blade residential fan. Both of these fans were very successful. Emerson manufactured six-blade residential fans long after all other major fan manufacturers had ceased production.

Nineteen-sixteen was an important year for Emerson. They introduced their Northwind line of inexpensive fans and perfected their oscillator. The first Northwind fan had a universal motor, two speeds, and was non-oscillating. The following year Emerson introduced a new line of 10" Northwinds that had three speeds and would oscillate. Northwind fans would be manufactured until 1929, and filled the niche for small, inexpensive fans all through the late teens and twenties.

The twenties was the first golden age of electric desk fan production. Emerson perfected their induction motor and oscillator following World War I and were ready to get into full gear manufacturing fans. In 1922 they introduced the Emerson Junior fan designed for home use. This line of fans sold very well for the company and was very popular with consumers.

When the Depression came, Emerson stopped manufacturing: junior fans, residential six-blade fans, and Northwind fans. They introduced the Seagull and Silver Swan in 1934 and the Seabreeze in 1936. The cross guard and bullet back were introduced in 1939 thus beginning the final phase of fan design for Emerson. The deep pitched overlapping blade also was introduced at this time.

The World War II years opened up many opportunities for Emerson. They received many government contracts for war-related manufactured goods. Following World War II the second golden age of electric desk fans blossomed. During the early post war years the demand for fans out stripped production. Emerson manufactured many high quality fans to meet this demand, but by the fifties, demand began to lag.

With the coming of affordable air conditioning in the late fifties, desk fans fell out of favor. By the sixties Emerson produced a much lower quality product than they had in the forties, but was still the leader in the industry. By the mid-sixties Emerson no longer manufactured desk fans and an important era had ended. (See page 65 for a list of Emerson numbers.)

Emerson began producing fans as soon as they started business. With the newly developed Meston Alternating Electric Motor, Emerson was off and running. The first three fans manufactured by Emerson had a tripod base attached to a large motor. The first of these being the Meston fan. There were several variations of Meston fans manufactured and they seemed to learn as they went along. These early fans were very fancy with bronze front and back motor plates. Meston fans came in four, five, and six winged blades and a guard was optional. Meston fans are extremely rare and very collectible. The next fan manufactured by Emerson (not pictured) had a pancake-type motor with a slide switch located at the rear of the motor. This Emerson fan had nickel-plated front and back plates and was very fancy. This was the first fan made by Emerson to have a guard as a standard feature. The tripod base had a black Japan finish with silver flowers as trim. By 1898 Emerson manufactured their final tripod fan that had a huge induction motor with fancy castings. This was the first fan to incorporate a brass guard and blade.

Left: ***Emerson***
Type: Meston
Blades: 12"
Date: 1891 – 1895

This was the first fan manufactured by Emerson. While this example is missing the blades, the fan could be purchased in four, five, or six winged blade configurations. The large Meston motor is attached to a cast-iron tripod base. The front and back plates are made of cast bronze and are very fancy. The base had ornate painted flower decoration. On top of the motor is a small lever that operates a variable speed control.

Grade I – $6,000.00*
Grade II – $4,500.00*
Grade III – $3,000.00*
Grade IV – $2,100.00*
Grade V – $1,200.00*
Grade VI – $300.00*

*This price reflects a fan with a blade. Add $500.00 for a guard.

Center: ***Emerson***
Type: F-2
Blades: 16"
Date: 1897 – 1899

This is an example of the third fan manufactured by Emerson. This fan has a very large induction motor with a fancy cast-iron motor housing. The fancy motor vents are trimmed in gold paint with flowers painted on the tripod base. What makes this fan extremely unusual is that it is the only 16" fan known to have been made by Emerson to have scalloped Parker blades. The guard is missing on this fan but was similar to the 12" fan from the same period.

Grade I – $3,600.00
Grade II – $2,700.00
Grade III – $1,800.00
Grade IV – $1,260.00
Grade V – $720.00
Grade VI – $180.00

Right: ***Emerson***
Type: F-1
Blades: 12"
Date: 1897 – 1899

This fan has a large ornate motor affixed to a tripod base. There are flowers painted on the base and gold trim around the ornate vent holes. There is a two speed rotary switch located at the back of the fan with a starting position. The blades and guard are made of brass. This was the first fan to use the scalloped Parker blades which became a standard feature on all better grade Emerson fans. This is the most commonly found Emerson tripod fan.

Grade I – $2,400.00
Grade II – $1,800.00
Grade III – $1,200.00
Grade IV – $840.00
Grade V – $480.00
Grade VI – $120.00

Motor Generator

.. .. or .. ..

Direct Current Converter,

.. .. for .. ..

Cautery and Other Uses.

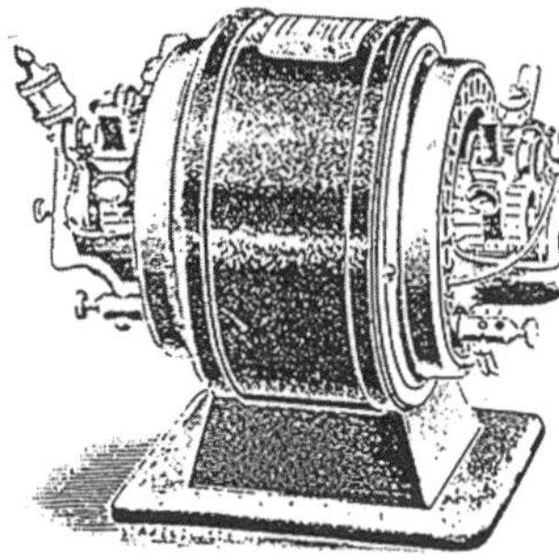

Manufactured for the Surgical Profession,
by us especially for

A. S. Aloe Company, St. Louis, Mo.

Small Fan Blower.

For Ventilation, Small Forges, Etc., Etc.

Patented Feb. 14, May 7 and 14, 1893, and Feb. 27, 1894.

9-inch Blower.............................price, $12 00
With Motor.............................. " 42 00

An illustration from an early Emerson catalog showing several uses for Emerson motors. The motor on the left is for direct current and has exposed oilers and brushes. The motor on the right is the Meston Alternating Current Motor used on Emerson's first fans. These motors had many applications and were marketed for such.

Our "3-inch" Alternating Current Motor.

For 15,000 and 16,000 Alternation Systems only.

FOR SMALL POWER PURPOSES.

Patented Feb. 14, March 7, March 14, 1893, and Feb. 27, 1894.

TYPE M. 2.

WE OFFER this Motor to the trade to meet a frequent call for a small Motor having power sufficient for dentists' grinding lathes and buff wheels, factory sewing machines, and laboratory work, such as static and similar machines, where a single high speed Motor will be effective.

We have recently put out numbers of these Motors, and they are giving universal satisfaction. They run free at about 4,000 turns per minute, and will load to best advantage at about 2,200.

They have regulating and reversing lever similar to Fan Motor; are neatly mounted on a low, heavy base. Their general appearance is decidedly neat.

An illustration from an early Emerson catalog showing a very small 3" Meston Alternating Current Motor. Note the small lever at the top right hand side of the motor. This was a speed regulator that would allow the motor to run at different speeds, even backwards. This feature was found on their early Meston fans and is unusual in the sense that it would allow the motor to run backwards.

Left: ***Emerson***
Type: EI-1
Blades: 12"
Date: 1898

Right: ***Emerson***
Type: PI 241
Blades: 12"
Date: 1899 – 1900

This shows the rear view of the last tripod fan made by Emerson. Note the very ornate casting with gold trim around the fancy vent holes. The initials EMECO cast around the bearing stand for Emerson Electric and Manufacturing Company. These initials were also used on their Meston fans. The rotary switch at the base of the back of the fan has a starting position and two speeds. Below the switch handle are two terminals to connect the fan to a power source. The tripod base has ornate painting which is typical of tripod fans. This example has five brass scalloped Parker blades attached to a cast-iron hub. This fan could be purchased in four, five, or six winged blades, four being the most common. The guard is made of formed brass wire and is very flimsy.

Grade I – $2,800.00
Grade II – $2,100.00
Grade III – $1,400.00
Grade IV – $980.00
Grade V – $560.00
Grade IV – $140.00

This fan is very important because it has several major design changes from the tripod fans. It is the first fan to incorporate the use of an ornate ribbed base in a cone configuration. This gives the fan a very modern look for 1899. The newly designed motor has been improved and made much smaller. The motor does not contain a centrifugal starting switch and had to be started manually by placing the slide switch lever in the start position and then moving it to the desired speed. The switch is very important in this fan and if it is missing it is a major loss. This fan also incorporates a newly designed guard that is far superior than the previous guard used on tripod fans. This open ring S-type guard became a standard feature on early Emerson fans until about World War I. The fan motor could be moved right or left on the base, but not up or down. If you wanted to mount this fan on the wall you had to purchase a special wall mount bracket. This was a very modern design fan for the period and established Emerson as a leader in fan manufacturing.

Grade I – $700.00
Grade II – $525.00
Grade III – $350.00
Grade IV – $245.00
Grade V – $140.00
Grade VI – $35.00

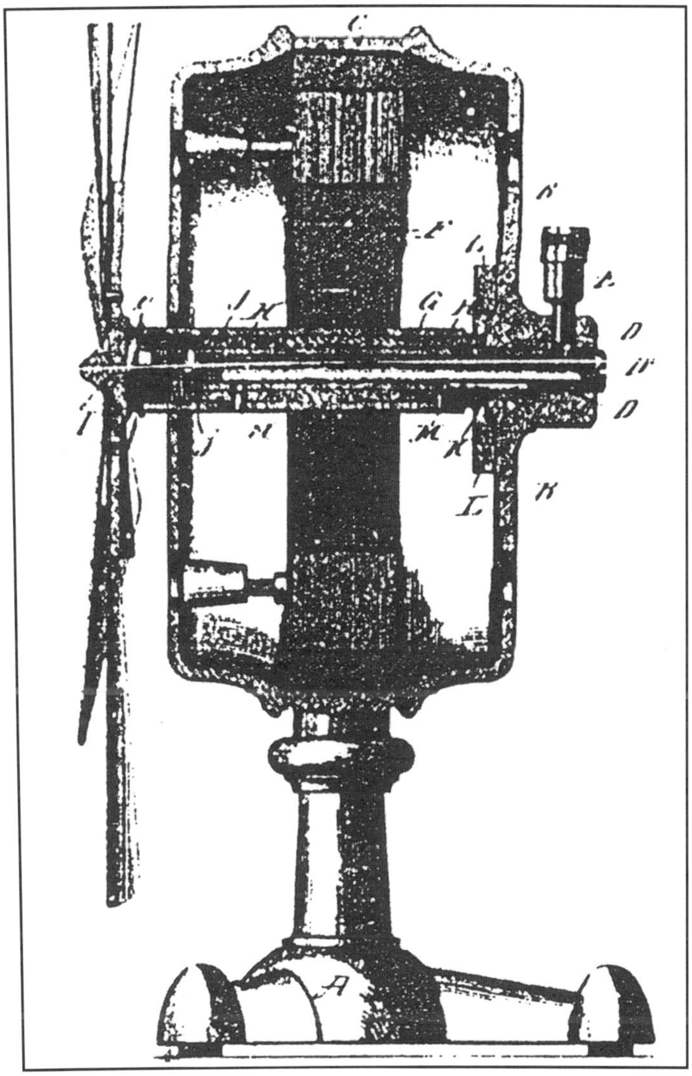

This is an illustration from a United States Patent office book from January 1899 showing a cross section of the Emerson single bearing motor. This cut out view shows the 1898 motor affixed to a tripod base. This tripod base is different than the Meston base because it allowed the fan motor to move from side to side, but not up down. A large oiler is located at the rear of the bearing for lubrication. This fan commonly had scalloped Parker blades, but this illustration has windmill type blades.

Emerson
Type: 1120
Blades: 16"
Date: 1903 – 1904

Emerson made large fans as soon as technology allowed them to do so, and these fans were primarily manufactured for commercial use. This example has an ornate motor with an internal starting switch. The ornate ribbed base has a two-speed slide switch. The large Parker blades are attached to a cast-iron hub. The guard is made of formed brass wire in the S- style with a large open center ring. The fan has a massive appearance.

Grade I – $1,500.00
Grade II – $1,125.00
Grade III – $750.00
Grade IV – $525.00
Grade V – $300.00
Grade VI – $75.00

Emerson
Type: 1020
Blades: 16"
Date: 1902

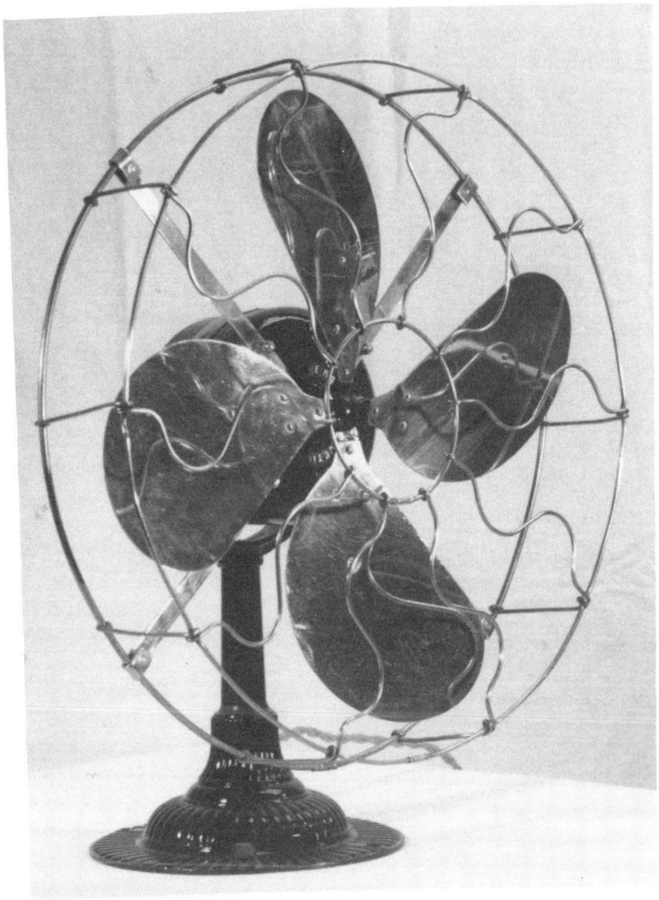

This fan was one of the first manufactured by Emerson to use a trunnion. A trunnion allowed the fan to be moved up or down and this was a major design development. Trunnions were commonly used on fans until about World War I. This fan has an ornate motor with an internal starting switch. The base is ornate and ribbed with a two-speed slide switch. The guard is made of formed brass wire in the open ring S-style. The Parker blades are brass and attached to a cast-iron hub. This is a very large fan for such an early date.

Grade I –	$1,300.00
Grade II –	$975.00
Grade III –	$650.00
Grade IV –	$455.00
Grade V –	$260.00
Grade VI –	$65.00

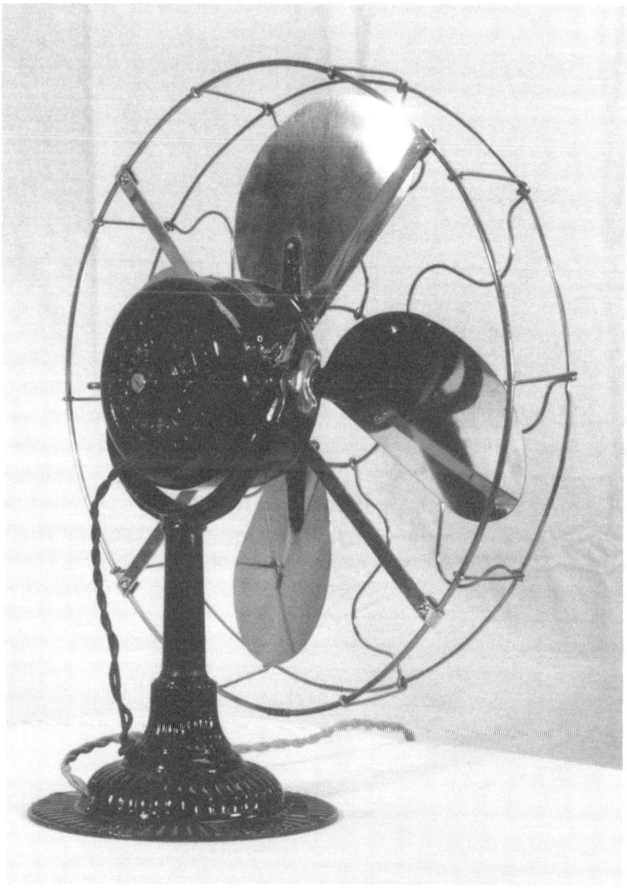

Left: ***Emerson Trojan***
Type: 54648
Blades: 16"
Date: 1912

Right: ***Emerson***
Type: 1820
Blades: 16"
Date: 1908 – 1909

The Trojan brand was an inexpensive line of desk and ceiling fans. They were manufactured from 1904 to 1912. These lower priced fans did not have as fancy castings as Emerson grade fans and never had Parker blades. This is one of the last Trojans produced and has an ornate motor with a smooth-step base. This fan has three speeds, which all other previous Trojans were two speed. The guard is one of the most important features that makes this fan unusual. It has a cast-brass Trojan badge which is a break away from the open ring guard so commonly used on Trojan fans. This fan incorporated Emerson's first lever-type oscillator that proved not very efficient. It is believed that this fan was made of parts from several other previously manufactured fans and placed on the newly designed smooth-step base.

Grade I – $700.00
Grade II – $525.00
Grade III – $350.00
Grade IV – $245.00
Grade V – $140.00
Grade VI – $35.00

This is a very large fan with an ornate motor affixed to the base by a trunnion. Note that on the top of the base is a ball. This is a joint of sorts that allows the fan to be wall mounted if desired. This ball became a standard feature on many 12" and 16" Emerson fans. This fan has large brass Parker blades attached to a cast-iron hub. The guard is massive and in the open S-ring design. This fan has two speeds and an internal starting switch in the motor.

Grade I – $900.00
Grade II – $675.00
Grade III – $450.00
Grade IV – $315.00
Grade V – $180.00
Grade VI – $45.00

Left: ***Emerson Trojan***
Type: 5110
Blades: 12"
Date: 1904 – 05

In the year of the St. Louis World's Fair, Emerson came out with one of its first Trojan desk fans. This fan is unusual due to the configuration of the motor and is also referred to as the Trojan-Pancake. The motor badge is placed at the center of the back of the motor, not a common feature of Trojan fans. The blades are windmill in style and are attached to a cast-iron hub. The base is ornate and ribbed with a large cast-iron trunnion. The guard is brass and has a small open ring configuration, different from Emerson grade fans of the period which had a large center ring. It has two speeds and is considered the most collectible of the Trojan line of fans.

Grade I – $700.00
Grade II – $525.00
Grade III – $350.00
Grade IV – $245.00
Grade V – $140.00
Grade VI – $35.00

Right: ***Emerson Trojan***
Type: 5610
Blades: 12"
Date: 1908 – 09

This is an early example of a six-blade residential fan. Residential fans were designed for home use and are not as common as the four-blade varieties. The motor is very massive, but plain. The air vents are large and more characteristic of General Electric fans of the period. The base is ornate and has the wall-mount ball where the trunnion and base meet. This fan has two speeds and a small center hole brass S-style guard characteristic of Trojan fans.

Grade I – $1,300.00
Grade II – $975.00
Grade III – $650.00
Grade IV – $455.00
Grade V – $260.00
Grade VI – $65.00

Left: ***Emerson Trojan***
Type: 53644
Blades: 8"
Date: 1911 – 12

All brass fans were popular with fan manufacturers around 1910, this is Emerson's example. This all brass, non-oscillating fan has 3 speeds and is very compact. It has a highly polished finish making it very attractive. The blades are made out of a single sheet of stamped brass and are unusual. This was the smallest Trojan fan ever manufactured by Emerson and is very collectible.

Grade I –	$700.00
Grade II –	$525.00
Grade III –	$350.00
Grade IV –	$245.00
Grade V –	$140.00
Grade VI –	$35.00

Center: ***Emerson Trojan***
Type: 5310
Blades: 12"
Date: 1906 – 09

This is the most commonly found Trojan fan. The cast-iron motor housing has large vent holes with an internal starting switch. The motor is affixed to the base by a trunnion without a wall-mount feature. This fan has two speeds which is a common characteristic on Trojan fans. The blades are the windmill type attached to a cast-iron spider. The guard is made of formed brass with a small-center S-ring.

Grade I –	$300.00
Grade II –	$225.00
Grade III –	$150.00
Grade IV –	$105.00
Grade V –	$60.00
Grade VI –	$15.00

Right: ***Emerson***
Type: 1500
Blades: 8"
Date: 1909

The 1500 was the first 8" fan made by Emerson. It was manufactured for home use almost exclusively and started a new trend in desk fan design. The blades are made from one sheet of formed brass in the scalloped Parker style. The guard is brass with a heavily cast-brass badge that is unique and is only found on this fan. The base is ornate ribbed with the motor attached by a trunnion. The ornate cast motor has an internal starting switch. This is the most collectible of the 8" fans made by Emerson, with the exception of the Type 15644 bedroom fan.

Grade I –	$900.00
Grade II –	$675.00
Grade III –	$450.00
Grade IV –	$315.00
Grade V –	$180.00
Grade VI –	$45.00

Left: ***Emerson***
Type: 2210
Blades: 12"
Date: 1909

Center: ***Emerson***
Type: 11666
Blades: 12"
Date: 1910

Right: ***Emerson***
Type: 1510
Blades: 12"
Date: 1906 – 09

This was the first oscillating fan made by Emerson and incorporates an ornate motor and ribbed base with a lever-type oscillator attached to the rear of the fan. The fan is moved back and forth by a series of levers and wheels. The operator could change the amount of oscillation by removing a set screw on the top guide wheel and placing the set screw in one of several locations. A lever attached to the wheel would move the fan back and forth. A very crude system that was abandoned in 1916 when Emerson developed their second oscillator. This fan has two bearings and uses a brass blade similar to those found on direct current fans of the period. The guard is the open ring style and the wall-mount joint at the top of the base has been modified so the fan does not have to be disabled to be placed in the wall-mount position.

Grade I – $700.00
Grade II – $525.00
Grade III – $350.00
Grade IV – $245.00
Grade V – $140.00
Grade VI – $35.00

In 1908 Emerson came out with its first residential fan which was a Type 1215. The fan pictured is an early prototype of this residential fan design. Residential fans ran slower than four-blade fans and have six sets of field coils to accommodate the larger load of the six blades. Because early residential fans had six sets of field coils, their massive motors almost eclipse the blades. The motor is very large and heavily cast and sits on a massive trunnion. The motor has an internal starting switch and is very quiet. Note the early cast-brass Emerson badge at the center of the guard, this was a change from the open ring style.

Grade I – $1,000.00
Grade II – $750.00
Grade III – $500.00
Grade IV – $350.00
Grade V – $200.00
Grade VI – $50.00

This was one of the last fans to use an open ring S guard and ornate ribbed base. The motor is ornately cast and contains a starting switch. This example has a handle on top of the motor. Handles were added to fans to keep people from picking up the fan by the guard. Handles became a standard feature on almost all Emerson 12" and 16" fans after 1914.

Grade I – $400.00
Grade II – $300.00
Grade III – $200.00
Grade IV – $140.00
Grade V – $80.00
Grade VI – $20.00

Left: ***Emerson***
Type: 12646
Blades: 12"
Date: 1910 – 11

Right: ***Emerson***
Type: 12666
Blades: 12"
Date: 1910 – 11

This is the second generation oscillating fan made by Emerson. The lever type oscillator used on the 2210 was somewhat modified for this fan, but was still rather inefficient and clumsy. This fan incorporated the single bearing motor which became a hallmark of Emerson fans. Emerson guaranteed that after lubricating your fan, excess oil would *never* run out of the bearing onto the spinning blades and spray the room. All fans of this period would commonly spray oil after lubrication, hence damaging furnishing and clothing. This was a revolutionary claim and brought Emerson many devoted customers. This has an ornate ribbed base with a wall-mount ball at the top of the base. The motor is ornate with an internal starting switch. The blades are scallop Parker blades attached to a cast-iron hub. The guard is made of formed brass wire with a cast-brass badge.

Grade I – $500.00
Grade II – $375.00
Grade III – $250.00
Grade IV – $175.00
Grade V – $100.00
Grade VI – $35.00

This fan was the first Emerson six-blade residential oscillator. The ornate motor is massive and very fancy almost eclipsing the blades. The motor has an internal starting switch. The base is cast-iron with ornate ribbing and a one-speed slide switch. The blades are scallop Parker in design and attached to a cast-iron hub. The blades and guard are brass with a cast-brass badge. This is a very quiet, efficient fan and very collectible.

Grade I – $900.00
Grade II – $675.00
Grade III – $450.00
Grade IV – $315.00
Grade V – $180.00
Grade VI – $45.00

Left: ***Emerson***
Type: 14644
Blades: 8"
Date: 1911 – 12

This fan has a smooth-step base and is a non-oscillator. The motor is on a trunnion and there is no internal starting switch in the motor. The fan has three speeds with brass scalloped Parker blades.

Grade I – $350.00
Grade II – $263.00
Grade III – $175.00
Grade IV – $123.00
Grade V – $70.00
Grade VI – $18.00

Center: ***Emerson***
Type: 11646
Blades: 12"
Date: 1910 – 11

A trunnion fan with an ornate ribbed base and motor with an internal starting switch. The blades are scalloped brass Parker blades attached to a cast-iron hub. There is a two-speed slide switch in the base.

Grade I – $500.00
Grade II – $375.00
Grade III – $250.00
Grade IV – $175.00
Grade V – $100.00
Grade VI – $35.00

Right: ***Emerson***
Type: 1010
Blades: 12"
Date: 1902 – 05

This is an example of an early trunnion fan. The ornate motor is attached to the base by a trunnion and has no wall-mount joint on the base. The guard is made of formed brass wire with a large open center ring. The brass scalloped Parker blades are attached to a cast-iron hub. A two-speed switch is located in the base of the fan. A very modern looking fan for 1902. This fan has an internal starting switch which was an innovation at the time.

Grade I – $600.00
Grade II – $450.00
Grade III – $300.00
Grade IV – $210.00
Grade V – $120.00
Grade VI – $30.00

Left: ***Emerson Trojan***
Type: 53646
Blades: 12"
Date: 1912

This is one of the very last Trojan fans made by Emerson. This example has an small open ring guard in the S-style made of formed brass wire. This guard possibly was left over stock from Emerson's ornate ribbed base fans that had fallen out of production. The motor is plane with no internal starting switch and is very similar to the 5310. Thc blades are brass windmill type attached to the motor by a cast-iron spider. This fan has both front and back bearings, a common characteristic of Trojan fans. The smooth-step base was new at this time, as well as the three-speed switch. A transitional fan made up of some old and new features; the last of the Trojan line.

Grade I – $700.00
Grade II – $525.00
Grade III – $350.00
Grade IV – $245.00
Grade V – $140.00
Grade VI – $35.00

Right: ***Emerson***
Type: 14666
Blades: 12"
Date: 1911

A residential fan with a large, plane motor attached to the base by a trunnion. The brass scalloped Parker blades are attached to the motor by a cast-iron hub. The motor is very large and quiet with three speeds. The guard made of formed brass wire with an Emerson cast-brass badge.

Grade I – $900.00
Grade II – $675.00
Grade III – $450.00
Grade IV – $315.00
Grade V – $180.00
Grade VI – $45.00

Left: ***Emerson***
Type: 16666
Blades: 12"
Date: 1912

Right: ***Emerson***
Type: 16646
Blades: 12"
Date: 1912

This is a residential oscillator with innovations such as a self-starting motor (no internal starting switch), newly designed smaller motor, smooth castings, and three speeds. The blades and guard are made of brass. Unfortunately, this fan entails the lever-type oscillator and is rather clumsy when in use. The fan is very quiet when operating. The wall-mount feature has been improved on the base, but the fan is still quite heavy for wall use.

Grade I – $600.00
Grade II – $450.00
Grade III – $300.00
Grade IV – $210.00
Grade V – $120.00
Grade VI – $80.00

One of the first modern looking Emerson oscillating fans, this has a smooth step base and plane cast-iron motor. There is a lever-type oscillator at the rear of the motor. The scalloped Parker blades are attached to a cast-iron hub. The guard is made of formed brass wire with a cast-brass badge. This fan has three speeds and looked like many Emerson oscillating fans which later followed. It is important to note that by the time Emerson had developed this fan the motor no longer contained an internal starting switch.

Grade I – $500.00
Grade II – $375.00
Grade III – $250.00
Grade IV – $175.00
Grade V – $100.00
Grade VI – $35.00

Both fans: ***Emerson***
Type: 19646
Blades: 12"
Date: 1914 – 1918

The 19000 series is a transitional group of fans which may have different characteristics. The fan on the left has a steel guard and a stamped steel hub with brass blades. The fan on the right has a brass guard, cast-brass badge, and a cast-iron hub with brass blades. At this time Emerson was changing from cast-iron hubs to stamped steel hubs and dropping the use of all brass guards. Besides these two differences, the fans are exactly the same. Note that each fan has a handle on top of the motor. It is hard to image how many brass guards were damaged by people picking up a fan by its flimsy brass guard. As World War I approached most major fan manufacturers began dropping brass guards and blades.

Brass Blades on Cast Hub/Brass Guard

Grade I	$300.00
Grade II	$225.00
Grade III	$150.00
Grade IV	$105.00
Grade V	$60.00
Grade VI	$15.00

Brass Blades/Steel Guard

Grade I	$150.00
Grade II	$112.00
Grade III	$75.00
Grade IV	$53.00
Grade V	$30.00
Grade VI	$8.00

Both fans: ***Emerson***
Type: 21646
Blades: 12"
Date: 1914 - 16

Pictured are the front and rear view of a fully developed Emerson desk fan. By 1914 Emerson began experimenting with a new oscillator that was finally patented in 1916. This new oscillator was much more efficient and easier to operate than the old lever type. This new oscillator contained a compact gear box located at the rear of the motor. A wheel located at the bottom of the gear box could be regulated to move the fan back and forth. This wheel had a small lever that stuck out from the wheel with a ball on the end of it. Notches in the wheel allowed for regulation. This levered wheel oscillator is referred to as the 1916 oscillator. It was modified again and used for decades on Emerson fans. Another innovation was the perfection of the wall-mount ball located at the top of the base of the stand. A collar was added that allowed the fan to be pivoted right or left as desired. This collar allowed the operator to set the fan motor in any horizontal direction and lock it in that position. This collar was used on Emerson fans for decades to come. Both these examples have brass blades and guard. The cast-iron hub was also still used, but was soon replaced by stamped steel.

Brass Blades on Cast Hub/Brass Guard	
Grade I –	$300.00
Grade II –	$225.00
Grade III –	$150.00
Grade IV –	$105.00
Grade V –	$60.00
Grade VI –	$15.00

Brass Blades/Steel Guard	
Grade I –	$150.00
Grade II –	$112.00
Grade III –	$75.00
Grade IV –	$53.00
Grade V –	$30.00
Grade VI –	$8.00

Left: ***Emerson***
Type: 24645
Blades: 9"
Date: 1917 – 19

This is a small version of the 21646 with several changes. The guard is brass, but the badge is not made of cast brass. The scalloped Parker brass blades are attached to the motor by a cast-iron hub. The motor is plane and has three speeds. It is a very desirable fan due to its small size and configuration.

Grade I – $400.00
Grade II – $300.00
Grade III – $200.00
Grade IV – $140.00
Grade V – $80.00
Grade VI – $20.00

Center: ***Emerson***
Type: 19645
Blades: 9"
Date: 1916 – 19

This example has a steel guard, but some examples may have a brass guard. The brass scalloped Parker blades are attached to the motor by a cast-iron hub. This fan has nice features and is very attractive. The thumb nuts are made of steel not brass, on this example.

Grade I – $300.00
Grade II – $225.00
Grade III – $150.00
Grade IV – $105.00
Grade V – $60.00
Grade VI – $15.00

Left: ***Emerson***
Type: 19045
Blades: 9"
Date: 1917 – 19

This is a direct current version of the 19645. This fan has a more compact motor than the alternating current version and is very powerful for its size. Direct current fans are often better constructed and more efficient than their alternating current counterparts. The motor is very modern looking for 1917. Direct current fans were very popular in the Northeast, Chicago, and St. Louis before World War I. Street cars and other city utilities used direct current power and it was readily available to urban consumers. The railroads also used many direct current fans in their operations. By World War II, direct current had fallen out of favor for cheaper, easier to transport, alternating current.

Grade I – $500.00
Grade II – $375.00
Grade III – $250.00
Grade IV – $175.00
Grade V – $100.00
Grade VI – $35.00

Left: ***Emerson Northwind***
Type: 44A
Blades: 8"
Date: 1916

By 1913 Emerson had dropped the Trojan line of fans and this had left a void in the lower priced market. In late 1915 Emerson engineers developed a new line of fans called Northwind. The first of these new inexpensive fans was the 44A. Several things set this fan apart from other fans of the period. First, it is painted gold, a very unusual color for such an early fan. Secondly, it had a universal motor so it could be operated both on alternating or direct currents. None of the major parts were brass and the blades were windmill type. The guard was small and only made out of four wires shaped in the S-style. The badge contained all the important information about the fan and there was no motor tag or other identification. Two speeds were kept as it had been on most of the old Trojan line of fans. The base was made out of cast pot metal and a modified trunnion held the motor to the base. This line of fans proved to be very successful for Emerson and they produced some form of Northwind until 1929.

Grade I – $90.00
Grade II – $68.00
Grade III – $45.00
Grade IV – $32.00
Grade V – $8.00
Grade VI – $5.00

Right: ***Northwind***
Type: 444A
Blades: 8"
Date: 1926

A similar fan to the 44A but not having a pot metal cast base. The fan looks more like the other Emerson fans of the twenties. The fan is finished in black with gold painted blades. It is two speeds with a universal AC/DC motor. These are very common fans and were produced in great quantities.

Grade I – $60.00
Grade II – $45.00
Grade III – $30.00
Grade IV – $21.00
Grade V – $12.00
Grade VI – $3.00

Left: ***Emerson Northwind***
Type: 450-C
Blades: 10"
Date: 1920 – 1921

Right: ***Northwind***
Type: 450 (no letter)
Blades: 10"
Date: 1929

With the success of the Northwind 44A, Emerson embarked on manufacturing a 10" Northwind series in 1917. Emerson produced this fan from 1917 until 1929 putting a letter after each consecutive year. A Northwind 450-D would have been made in 1921. A fan ending with just the number 450 would have been manufactured in 1929 at the end of the series. Not all letters were used, so you must be careful in calculating age. The first Northwind 450-A came out in 1917, and the last 450-L came out in 1928. The 1929 Northwind 450 had no letter and ended the series. The letter "I" was never used. These fans had several general characteristics such as all being oscillators, having three speeds, windmill-type painted blades, and a four "S" wire guard. It should be noted that each year the fan was slightly modified. Some examples have gold painted blades, others black. The oscillator wheel may be the 1916 type, continuous or standard. Some early models have a ball at the top of the base for wall mount purposes. Each year was slightly different. These are very common fans and were manufactured in great quantities.

Grade I –	$90.00	Grade IV –	$32.00
Grade II –	$68.00	Grade V –	$18.00
Grade III –	$45.00	Grade VI –	$5.00

Left: ***Emerson***
Type: Emerson Junior
Blades: 8"
Date: 1923

Right: ***Emerson Northwind***
Type: 55
Blades: 8"
Date: 1929

In 1922, Emerson came out with the Emerson Junior line of fans. These fans were to meet the growing need of household consumers following World War I. The first Emerson Junior was a 9" plug-in fan that met with great success. The following year they come out with an 8" oscillator with a slide switch in the base. These 8" fans were not produced in the quantities as the 9" and 10" juniors, thus the 8" are more collectible. The base and motor housing are cast-iron with the front motor plate being made out of stamped steel. The blades are painted gold and give the appearance of brass, but no junior fans had brass blades. Emerson was still using brass on some of their 12" and 16" fans at this time and were painting them gold. These painted blades may have made consumers believe these smaller fans were brass painted blades also. These are excellent fans of the highest quality and millions were sold at the time.

Grade I – $110.00
Grade II – $83.00
Grade III – $55.00
Grade IV – $39.00
Grade V – $22.00
Grade VI – $6.00

This is the last in a line of small Northwind fans made by Emerson. By the end of production, the guard had been modified from four to six wires. The blades are of the windmill type and the base and motor housing are stamped steel, not cast-iron. The traditional slide switch has been replaced with a toggle switch. By 1929 the universal AC/DC motor had fallen out of favor due in part to the disruption of radio reception and the limited use of direct current electricity. These fans came in bright red and green as well as cream.

Grade I – $50.00
Grade II – $38.00
Grade III – $25.00
Grade IV – $18.00
Grade V – $10.00
Grade VI – $3.00

Right: ***Emerson Junior***
Type: Junior
Blades: 10"
Date: c.1925 – 1928

The 10" Emerson Junior became the standard home use fan made by Emerson in the later 1920s. This fan has a cast-iron base and motor housing with a stamped steel motor front plate. The blades are stamped steel in a scalloped Parker-style painted gold. All junior fans of this period are one speed with a slide switch in the base. An easy way to date junior fans are by the number of holes in the base. The fans with four holes are from the early 20s, the fans with three holes are later models. This rule applies for all 8", 9", and 10" models.

Grade I – $70.00
Grade II – $53.00
Grade III – $35.00
Grade IV – $25.00
Grade V – $14.00
Grade VI – $3.00

Center: ***Emerson Junior***
Type: Junior
Blades: 9"
Date: c.1925 – 1929

The 9" Emerson Junior was one of the most popular fans ever manufactured by Emerson. The example pictured has four holes in the base with the slide switch to the side, not in front. The earlier fans had a red, gold, and black badge with all the information about the fan found on the badge. These fans had no information badges on top of the motor.

Grade I – $90.00
Grade II – $68.00
Grade III – $45.00
Grade IV – $32.00
Grade V – $18.00
Grade VI – $5.00

Right: ***Emerson Junior***
Type: Junior
Blades: 10"
Date: c.1925 – 1928

An example of an non-oscillating junior fan. A later example with three holes in the base with the slide switch in the front of the base.

Grade I – $50.00
Grade II – $38.00
Grade III – $25.00
Grade IV – $18.00
Grade V – $10.00
Grade VI – $3.00

Emerson Fans Guaranteed 5 Years

CATALOG No. 4037

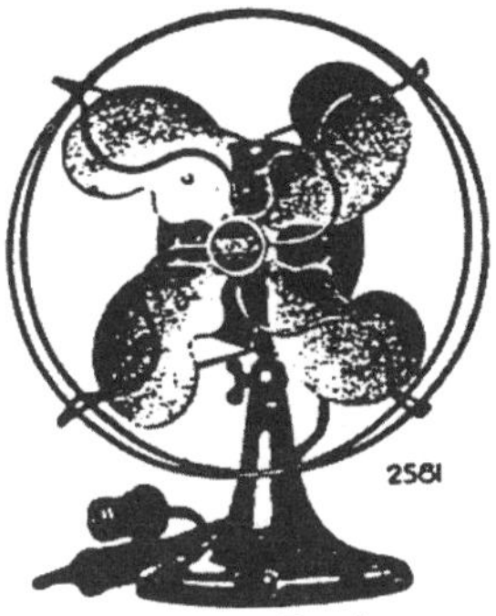

Emerson Junior
9-inch Non-oscillator

Both types are adjustable
for wall mounting

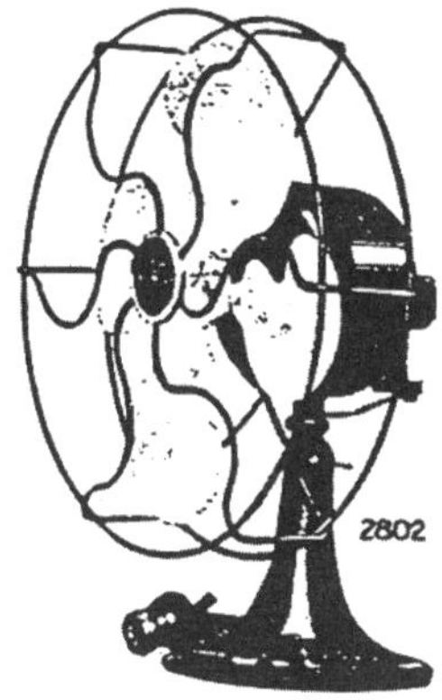

Emerson Junior
10-inch Oscillator

Emerson Junior 9-inch and 10-inch Fans

for Alternating Current

THE Emerson Junior fans have big sales features, and yet they sell at popular prices. They are built of the same high grade material and workmanship as the larger Emerson fans, and they are also guaranteed for 5 years.

The two sizes answer every need for a small, quality fan for general purposes, and the large volume of sales proves their ever increasing popularity.

Specifications

BLADES—Parker design, dull brass finish.

SHAFT—¼-inch hardened steel shaft with oil tight, dust-proof bearing.

BRACKET BASE—Hinged desk-bracket base, 9-inch fan has four large rubbers on base, 10-inch fan felt covered base.

SWITCH—In base—single speed.

MOTOR—Induction type, no brushes or commutator.

GUARD—Double ring, steel, welded joints.

FINISH—Guard, motor and base, black enamel finish.

CORD AND PLUG—Cord and separable plug furnished complete with every fan.

OSCILLATOR—10-inch fan has two ranges of oscillation 45° and 90° or the non-oscillating position secured by means of the adjusting disc.

—Guaranteed for 5 years—

Codewords—Data and List Prices

Codeword	Volts	Cycles	Size Fan Inches	No. Blades	Style	Speed and Wattage	Net Lbs.	Packed Lbs.	List Price Each
Soteb	110	60	9	4	Hinged Non-oscillating	1600 rpm 45 watts	9	15	$10.00
Sylok	110	60	10	4	Hinged Oscillating	1600 rpm 45 watts	12	17	16.50

Regularly carried in stock at St. Louis, New York, Chicago and by conveniently located distributors.

Emerson Junior fans listed for 110 volts 60 cycles, will operate satisfactorily on 104-120 volts, 60 cycles and 104-115 volts, 50 cycles.

On 50 cycles speed will be approximately 15 % lower.

Emerson Junior fans cannot be supplied for other kinds of current.

Speeds, wattages and weights are approximate. All data subject to change.

An illustration from a 1927 Emerson catalog advertising Emerson Junior fans. These fans came in 8", 9", and 10" sizes. The 9" and 10" could be purchased in oscillating and non-oscillating models. The Emerson Junior line was a major seller for Emerson and was an excellent fan. These fans had the single bearing construction which was a hallmark of Emerson fans. It is interesting to note that these fans could not be purchased in direct current models.

Left: ***Emerson***
Type: 29645
Blades: 9"
Date: c.1922 – 1925

One of the very last fans made by Emerson to use brass blades. The base looks too high for the fan and makes for a strange configuration. The badge on this example is characteristic of an earlier model. The base and motor housing are cast-iron. The guard is made of formed steel with a brass and black badge. The fan has three speeds and will oscillate.

Grade I – $130.00
Grade II – $98.00
Grade III – $65.00
Grade IV – $45.00
Grade V – $26.00
Grade VI – $7.00

Right: ***Emerson***
Type: 29646
Blades: 12"
Date: 1922 – 1925

Millions of these fans were manufactured in the twenties and were one of Emerson's most popular fans. The blades are brass painted gold with a spot welded steel guard. The base and motor housing are cast-iron with a three-speed slide switch in the base. A handle is placed on the top of the motor for easy transport. This fan is an oscillator, but they were manufactured in non-oscillating models.

Grade I – $90.00
Grade II – $68.00
Grade III – $45.00
Grade IV – $32.00
Grade V – $18.00
Grade VI – $5.00

Left: ***Emerson***
Type: 73668
Blades: 16"
Dute: 1928 – 1929

Right: ***Emerson***
Type: 27666
Blades: 16"
Date: 1919 – 1922

A very large six-blade residential type fan for industrial use. This example has the modified Improved Parker fan blade used on later 16" Emerson fans. This fan has the later Pyramid badge found on fans made in the later twenties and early thirties. The blades are brass painted gold, the guard is made of heavy spot welded steel. The base and motor housing are cast-iron and quite heavy. The handle on top of the motor is a necessity with this fan due to its large size. It has three speeds and will oscillate.

Grade I – $200.00
Grade II – $150.00
Grade III – $100.00
Grade IV – $70.00
Grade V – $40.00
Grade VI – $10.00

A large six-blade residential type fan with traditional highly polished Parker blades. The base and motor housing are made of cast-iron. This fan has three speeds and will oscillate. This was the last model to have the large motor. There is a handle on top for easy transport.

Grade I – $300.00
Grade II – $225.00
Grade III – $150.00
Grade IV – $105.00
Grade V – $60.00
Grade VI – $15.00

Left: ***Emerson***
Type: 27046
Blades: 12"
Date: 1919 – 1923

A direct current fan with highly polished brass scalloped Parker blades. The round ball motor has front brushes which was characteristic of direct current fans of the period. The base and motor housing are made of cast-iron. The low speed is very slow and characteristic of direct current fans made by Emerson. The motor is very quiet and moves a lot of air when operating.

Grade I – $350.00
Grade II – $263.00
Grade III – $175.00
Grade IV – $123.00
Grade V – $70.00
Grade VI – $18.00

Right: ***Emerson***
Type: 27048
Blades: 16"
Date: 1919 – 1924

A large direct current fan with highly polished Parker blades. This also has a large round ball motor with front end brushes. The base and motor housing are made of cast iron. This fan blows a large amount of air, will oscillate quickly, and was designed primarily for industrial use.

Grade I – $450.00
Grade II – $338.00
Grade III – $225.00
Grade IV – $158.00
Grade V – $90.00
Grade VI – $23.00

Left: ***Emerson***
Type: 24648
Blades: 16"
Date: 1917 – 19

Right: ***Emerson***
Type: 24666
Blades: 12"
Date: 1917 – 18

The 24000 series was another transitional group of fans made around World War I. It was the first series to use the new smooth-cone base which became the standard on all Emerson fans. The base was open at the bottom like the smooth-step base it replaced and rested on six small rubber grommets set into the base. By the time this fan was manufactured the wall mount ball at the top of the motor had been fully developed. There were two kinds of bases produced at this time, a standard size (pictured) and a smaller size. It seems that Emerson was experimenting with this newly designed base and wanted to see which base would work best. The smaller base was rather unstable and tended to allow the fan to fall when disturbed. The collector may find any 24000 series with either; a brass or steel guard, cast-iron or stamped-steel fan blade hub, a 1916 oscillator wheel or later oscillator wheel. These differences will effect the value of each fan. An example with a new oscillator, stamped-steel fan hub and steel guard would be less than an example with a brass guard, cast-iron hub, and 1916 oscillator. It should be noted that this was the last series to use the large motor. The next series had a more modern compact motor.

Brass Blades on Cast Hub/ Brass Guard		Brass Blades/ Steel Guard	
Grade I –	$500.00	Grade I –	$350.00
Grade II –	$375.00	Grade II –	$263.00
Grade III –	$250.00	Grade III –	$175.00
Grade IV –	$175.00	Grade IV –	$123.00
Grade V –	$100.00	Grade V –	$70.00
Grade VI –	$35.00	Grade VI –	$18.00

This was the last 12" residential six-blade fan to use a large motor. These are very massive fans. Some examples will have a brass guard and blades attached to a cast-iron hub, others will have the smaller base are rather unstable and should be handled with care. These are very quiet and efficient fans.

Brass Blades on Cast Hub/ Brass Guard		Brass Blades/ Steel Guard	
Grade I –	$700.00	Grade I –	$400.00
Grade II –	$525.00	Grade II –	$300.00
Grade III –	$350.00	Grade III –	$200.00
Grade IV –	$245.00	Grade IV –	$140.00
Grade V –	$140.00	Grade V –	$80.00
Grade VI –	$35.00	Grade VI –	$20.00

Left: ***Emerson***
Type: 71045
Blades: 10"
Date: 1923

Emerson would finish any fan they manufactured in a color of the buyer's choice and this is an example. The first fan is a direct current oscillator finished in dark brown with a black brass Pyramid badge trimmed in brown. Note that the brush caps are in the front of the motor. The fan has a 3-speed switch in the base with scalloped Parker steel blades.

Grade I –	$300.00
Grade II –	$225.00
Grade III –	$150.00
Grade IV –	$105.00
Grade V –	$60.00
Grade VI –	$15.00

Left: ***Emerson***
Type: 79045
Blades: 10"
Date: unknown

This is another direct current fan that is finished in a light cream color. Note the badge. Emerson always painted the badge to match the custom finish of the fan. If you find a fan painted a color other than black and the badge is black, most likely the fan was not painted at the factory. Paint in lighter colors tended to become brittle and chip after a long period of time. This did not seem to be the case with the gloss black finish.

Grade I –	$400.00
Grade II –	$300.00
Grade III –	$200.00
Grade IV –	$140.00
Grade V –	$80.00
Grade VI –	$20.00

Left: ***Emerson***
Type: Emerson B-Junior
Blades: 10"
Date: 1936

The Depression changed fan design and marketing strategy greatly. This group of fans represents a newly designed line of products for home and commercial use. The motor has been redesigned and made more efficient. The single bearing design for 10" fans has been discontinued. The body of the motor is still made of cast-iron, as well as the base. A single speed slide switch is placed in front of the base with "on and off" cast underneath the slide. The information about the fan is found on the front badge which is made of stamped steel instead of brass. The guard is held to the motor by four bolts instead of the two found on older juniors. These fans were produced in great numbers and were a successful product for Emerson.

Grade I – $50.00
Grade II – $38.00
Grade III – $25.00
Grade IV – $18.00
Grade V – $10.00
Grade VI – $3.00

Center: ***Emerson***
Type: Emerson B-Junior
Blades: 10"
Date: 1936

This example has two speeds making it a more deluxe model. The base is winder than the single speed versions and the badge is brass and black, not silver and black. It has the newly designed motor and scalloped Parker blades finished in flat black. This example will oscillate. The two speeds makes this fan quite unusual.

Grade I – $100.00
Grade II – $75.00
Grade III – $50.00
Grade IV – $35.00
Grade V – $20.00
Grade VI – $5.00

Right: ***Emerson***
Type: 2250-C
Blades: 10"
Date: 1938

A later version of the 1936 Emerson B-Junior. It seems that by 1938 Emerson dropped the B-Junior name and began making this fan as a full grade Emerson. A version of this fan was manufactured until about 1950.

Grade I – $50.00
Grade II – $38.00
Grade III – $25.00
Grade IV – $18.00
Grade V – $10.00
Grade VI – $3.00

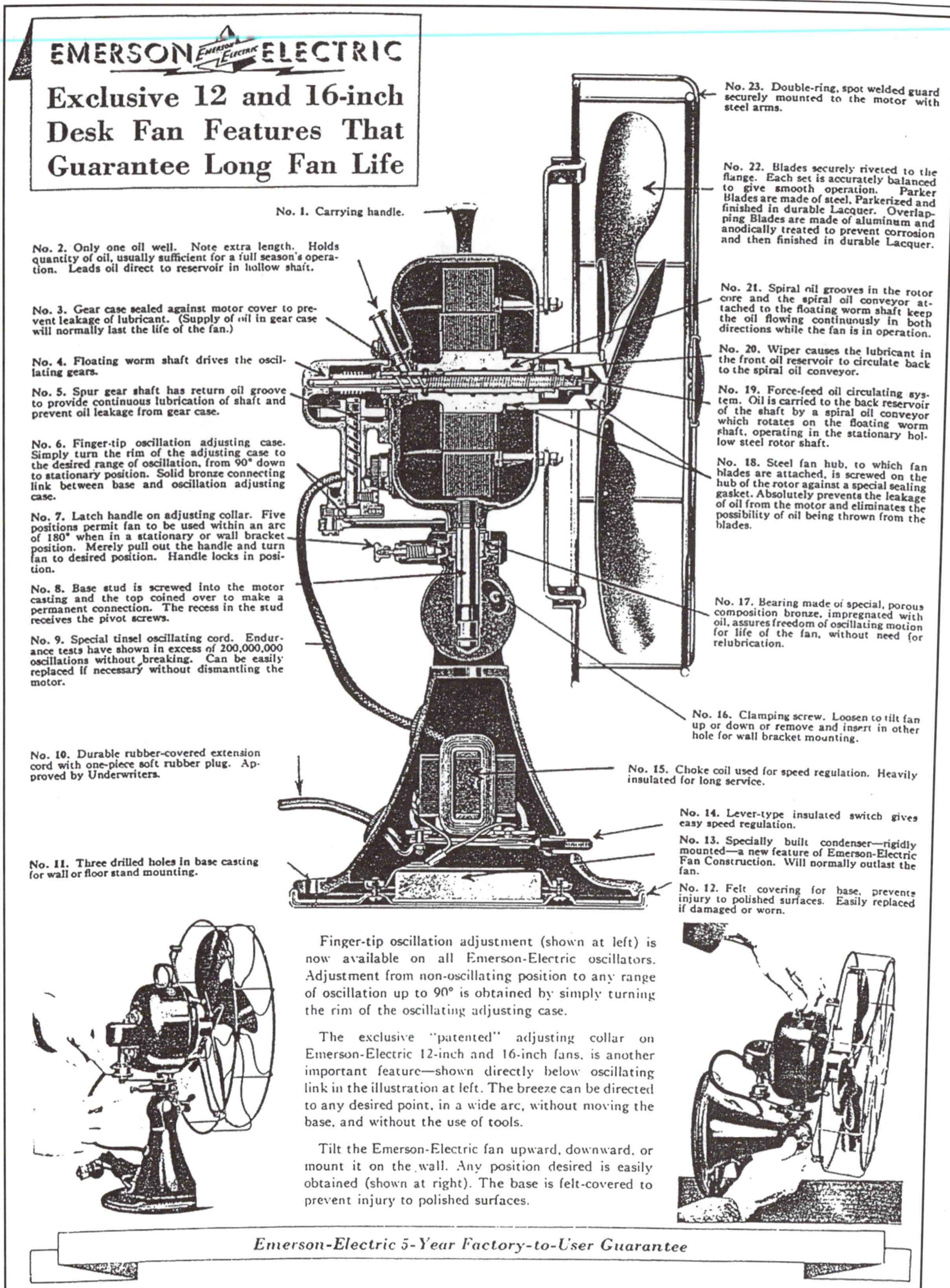

EMERSON ELECTRIC

Exclusive 12 and 16-inch Desk Fan Features That Guarantee Long Fan Life

No. 1. Carrying handle.

No. 2. Only one oil well. Note extra length. Holds quantity of oil, usually sufficient for a full season's operation. Leads oil direct to reservoir in hollow shaft.

No. 3. Gear case sealed against motor cover to prevent leakage of lubricant. (Supply of oil in gear case will normally last the life of the fan.)

No. 4. Floating worm shaft drives the oscillating gears.

No. 5. Spur gear shaft has return oil groove to provide continuous lubrication of shaft and prevent oil leakage from gear case.

No. 6. Finger-tip oscillation adjusting case. Simply turn the rim of the adjusting case to the desired range of oscillation, from 90° down to stationary position. Solid bronze connecting link between base and oscillation adjusting case.

No. 7. Latch handle on adjusting collar. Five positions permit fan to be used within an arc of 180° when in a stationary or wall bracket position. Merely pull out the handle and turn fan to desired position. Handle locks in position.

No. 8. Base stud is screwed into the motor casting and the top coined over to make a permanent connection. The recess in the stud receives the pivot screws.

No. 9. Special tinsel oscillating cord. Endurance tests have shown in excess of 200,000,000 oscillations without breaking. Can be easily replaced if necessary without dismantling the motor.

No. 10. Durable rubber-covered extension cord with one-piece soft rubber plug. Approved by Underwriters.

No. 11. Three drilled holes in base casting for wall or floor stand mounting.

No. 12. Felt covering for base, prevents injury to polished surfaces. Easily replaced if damaged or worn.

No. 13. Specially built condenser—rigidly mounted—a new feature of Emerson-Electric Fan Construction. Will normally outlast the fan.

No. 14. Lever-type insulated switch gives easy speed regulation.

No. 15. Choke coil used for speed regulation. Heavily insulated for long service.

No. 16. Clamping screw. Loosen to tilt fan up or down or remove and insert in other hole for wall bracket mounting.

No. 17. Bearing made of special, porous composition bronze, impregnated with oil, assures freedom of oscillating motion for life of the fan, without need for relubrication.

No. 18. Steel fan hub, to which fan blades are attached, is screwed on the hub of the rotor against a special sealing gasket. Absolutely prevents the leakage of oil from the motor and eliminates the possibility of oil being thrown from the blades.

No. 19. Force-feed oil circulating system. Oil is carried to the back reservoir of the shaft by a spiral oil conveyor which rotates on the floating worm shaft, operating in the stationary hollow steel rotor shaft.

No. 20. Wiper causes the lubricant in the front oil reservoir to circulate back to the spiral oil conveyor.

No. 21. Spiral oil grooves in the rotor core and the spiral oil conveyor attached to the floating worm shaft keep the oil flowing continuously in both directions while the fan is in operation.

No. 22. Blades securely riveted to the flange. Each set is accurately balanced to give smooth operation. Parker Blades are made of steel, Parkerized and finished in durable Lacquer. Overlapping Blades are made of aluminum and anodically treated to prevent corrosion and then finished in durable Lacquer.

No. 23. Double-ring, spot welded guard securely mounted to the motor with steel arms.

Finger-tip oscillation adjustment (shown at left) is now available on all Emerson-Electric oscillators. Adjustment from non-oscillating position to any range of oscillation up to 90° is obtained by simply turning the rim of the oscillating adjusting case.

The exclusive "patented" adjusting collar on Emerson-Electric 12-inch and 16-inch fans, is another important feature—shown directly below oscillating link in the illustration at left. The breeze can be directed to any desired point, in a wide arc, without moving the base, and without the use of tools.

Tilt the Emerson-Electric fan upward, downward, or mount it on the wall. Any position desired is easily obtained (shown at right). The base is felt-covered to prevent injury to polished surfaces.

Emerson-Electric 5-Year Factory-to-User Guarantee

An illustration from a 1937 fan catalog of a standard Emerson desk fan with scalloped blades. These fans were manufactured in large numbers and were very durable. The early prototype of these fans had brass blades painted gold but by the late twenties the brass had been replaced by steel. These fans were of the highest quality and sold well for Emerson.

Left and Center: ***Emerson***
Type: 71666
Blades: 12"
Date: 1923 – 1932

Right: ***Emerson***
Type: 71666 FG
Blades: 12"
Date: 1923 – 1932

In 1923 Emerson came out with a line of residential fans for home use. These fans had newly designed smaller more efficient motors. The early examples (left) had brass blades painted gold with a brass and black badge. All examples had steel guards. Brass guards had been completely discontinued from production for easier-to-produce steel guards. The center picture shows a late example of this type fan. The blades are painted steel in the scalloped Parker style finished in a flat black finish. The newer fans have the Pyramid badge which came out in the middle 1920s. The motor badge may be slightly different on the later models but will always have the same type number.

Brass Blades / Steel Guard
Grade I – $200.00
Grade II – $300.00
Grade III – $100.00
Grade IV – $140.00
Grade V – $80.00
Grade VI – $20.00

Steel Blades / Steel Guard
Grade I – $100.00
Grade II – $75.00
Grade III – $50.00
Grade IV – $35.00
Grade V – $20.00
Grade VI – $5.00

The example on the right is a residential six-blade fan finished in French Gray. French Gray was a special finish that could be ordered from the factory. Emerson produced fan literature all through the twenties and early thirties touting this finish. Unfortunately, the paint was flawed and did not hold up as well as the gloss black finish. All of the examples I have seen have had chips and discoloration throughout the finish. These are very handsome and attractive fans and were very popular among consumers.

Grade I – $350.00
Grade II – $263.00
Grade III – $175.00
Grade IV – $123.00
Grade V – $70.00
Grade VI – $18.00

Emerson Fans Guaranteed 5 Years

CATALOG Nº 4037

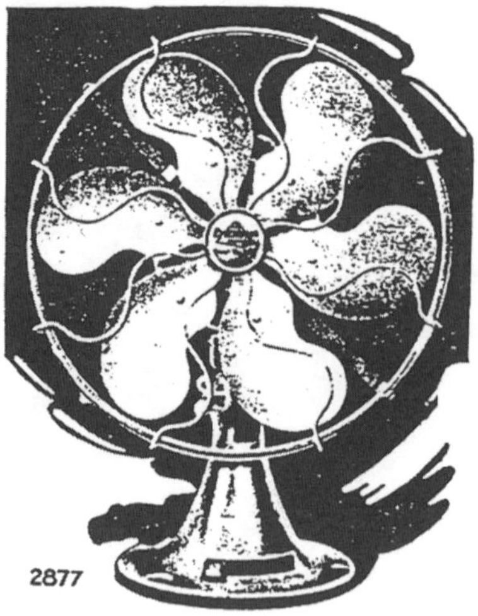

2877

Emerson 12-inch French Gray Finish Fan Slow-Speed Oscillator

for Alternating Current

THIS is the Emerson fan intended particularly for the home, although it is also used by physicians and dentists in their offices and reception rooms. The French Gray finish harmonizes with practically any color scheme, and for this reason is well adapted for use in any room in the home. Many hospitals are also adopting this fan with French Gray finish as standard equipment.

Specifications

BLADES—Parker design, brass.

SHAFT—½-inch hardened steel shaft and oil tight dust-proof bearing.

BASE—Hinged desk-bracket base for wall mounting, base felt covered.

ADJUSTING COLLAR—Swivel adjusting collar below motor permits range of adjustment to 180°.

OSCILLATOR—Range of oscillation adjustable to 90° from any position set by adjusting collar.

SWITCH—Three speeds, switch in base.

MOTOR—Induction type—no brushes or commutator.

GUARD—Steel, double-ring, welded joints.

FINISH—French Gray "Duco" over all.

CORD AND PLUG—Every fan complete with cord and separable plug.

HANDLE—Large carrying handle.

—Guaranteed for 5 years—

Codeword—Data and List Price

Codeword	Volts	Cycles	Size Fan Inches	No. Blades	Style	Type No.	Fast Speed	Medium Speed	Slow Speed	Net Lbs.	Packed Lbs.	List Price Each
Sotir	110	60	12	6	Oscillating	71666F.G.	1050 rpm 70 watts	950 rpm 50 watts	850 rpm 45 watts	20	38	$35.00

Regularly carried in stock at St. Louis, New York, Chicago and by conveniently located distributors.

Will operate satisfactorily on 104-120 volts.

Speeds, wattages and weights are approximate.

All data subject to change.

An illustration of a six-blade residential oscillator in a French Gray finish. This light colored fan was advertised for home or special use. This picture is from a 1929 catalog. Emerson made six-blade residential fans long after most manufactures had dropped production. Unfortunately this finish was not as durable as the gloss black finish and did not hold up well to the test of time.

Left: ***Emerson***
Type: 29648
Blades: 16"
Date: 1922 – 25

Right: ***Emerson***
Type: 73638
Blades: 16"
Date: 1932

In the early twenties Emerson came out with a newly designed 16" Parker blades called the Parker Improved blade. These blades were narrower than their older counterparts and are almost never made of brass. These fans were manufactured primarily for industrial use and mass produced from about 1921 through about the mid-thirties. Note that some of the examples have twisted guard brackets which was an attempt to quiet blade noise. The motor badge and guard badge are painted steel and are never brass. This example has a guard badge in the newer Pyramid style with the "Built to last" motto.

Grade I –	$150.00
Grade II –	$113.00
Grade III –	$75.00
Grade IV –	$53.00
Grade V –	$30.00
Grade VI –	$8.00

Another example of a large industrial fan with Parker Improved blades. The Pyramid badge is less detailed with motto, "Built to last" missing. The guard brackets are not twisted and look rather flimsy.

Grade I –	$150.00
Grade II –	$113.00
Grade III –	$75.00
Grade IV –	$53.00
Grade V –	$30.00
Grade VI –	$8.00

EMERSON Chromium Finish Fans...

These Fans Are Coming into Wider Use Every Year. The Sparkling Chromium Finish Adds a Touch of Beauty and Refinement to the Most Elaborate Furnishings.

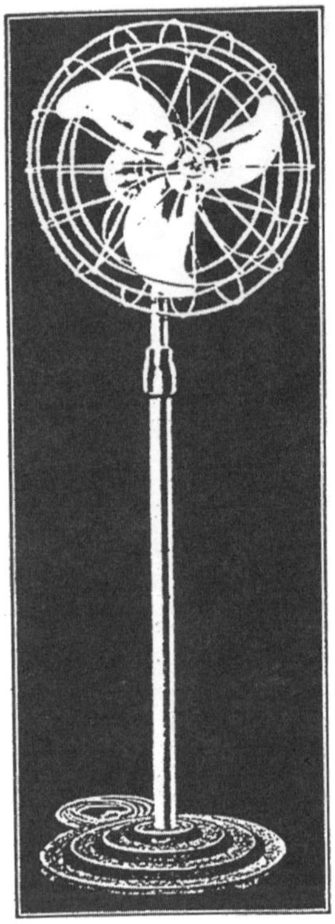

Type S60WA Air Circulator on Adjustable Floor Column Mounting One-Year Guarantee

Emerson Chromium finished fans are ideal for finely furnished stores, club-rooms, restaurants, theatres, and similar public places.

Emerson oscillators are available in three sizes for alternating current and two sizes for direct current. Detail specifications and complete performance information are given on pages 10, 11, 12 and 13.

Emerson Air Circulators and Accessories are also available from stock, in five types and four styles of mountings. The gleaming, polished aluminum blades and chromium-plated motor covers blend beautifully with the satin-finish, chromium-plated guard and aluminum lacquer finish on the motor field ring. For Air Circulator specifications and performance information turn to pages 14 and 15.

Finish Specifications Only

DESK FANS

FINISH—Polished chromium plate over all parts.

AIR CIRCULATORS

MOTOR FINISH: Field ring and mounting yoke attachment finished in aluminum lacquer. Front end cover and back enclosing shell polished chromium plate.

BLADE FINISH: Type S60WC—polished chromium plate, other types highly polished aluminum.

GUARD FINISH: Satin-finish chromium plate.

Air Circulator Mounting Accessories

CEILING MOUNTING FITTINGS: Motor-socket and ceiling shell finished in polished chromium plate. (No hanger pipe regularly furnished. See note at bottom of page for list price.)

FLOOR COLUMN MOUNTING: Base finished in black wrinkle, coupling in baked black japan, column and motor-socket in polished chromium plate.

COUNTER COLUMN MOUNTING: Column, base and motor-socket finished in polished chromium plate.

WALL BRACKET MOUNTING: Bracket, stud and motor-socket finished in polished chromium plate.

Type 73646-AK 12-inch Oscillator 5-Year Guarantee

SIZE	No. BLADES	VOLTS	CURRENT	SPEEDS	PERFORMANCE	WEIGHT		TYPE No.	CODE WORD	LIST PRICE EACH
						NET	PKD.			
10"	4	110	60 cy.	1 sp. osc.	For complete performance information on these fans see pages 10, 13 and 14.	11	16	2250-B	SEDUN	$18.95
12"	4	110	60 cy.	3 sp. osc.		21	31	73646-AK	SARAK	33.50
16"	4	110	60 cy.	3 sp. osc.		23	38	73648-AK	SARON	40.50
12"	4	115	D. C.	3 sp. osc.		23	38	75046	TIPON	37.00
16"	4	115	D. C.	3 sp. osc.		25	42	75048	TIVOT	43.50
24"	4	110	60 cy.	1 sp.		36	59	S60WC	KOCOP	48.75
24"	3	110	60 cy.	2 sp.		37	60	S60WA	KLANT	55.25
30"	3	110	60 cy.	2 sp.		41	62	S60WB	KHOTM	67.75
24"	3	115	D. C.	1 sp.		38	61	D60TL	TESOF	63.75
30"	3	115	D. C.	1 sp.		42	63	D60TM	TESAB	76.25
Ceiling Mounting Fittings for Air Circulator (no hanger pipe)						5	7	....	KNIPS	3.75
Adjustable Floor Column Mounting for Air Circulator, Min. 4'11", Max. 8'4"						58	74	....	KOGAP	19.00
Counter Column Mounting for Air Circulator, 1'8" base to center of fan						33	41	....	KNOLP	13.50
Wall Bracket Mounting for Air Circulator, 1'2" extension from wall						15	25	....	KLEMT	12.00

All chrome fans became popular in the late twenties and thirties. Emerson would chrome plate any of their products for an added cost. This is an illustration from a 1937 fan catalog advertising chromium finished fans. These fans were touted as being stylish and very modern looking. Air circulators were often finished in chromium as a standard feature but desk fan examples are less common.

In 1934 Emerson introduced their Seagull line of fans. This is an illustration from a 1934 catalog. It states the Seagull is "not just another fan...it's the crowing achievement of Emerson's long history." That could not have been farther from the truth. The Seagull had eye appeal, but no "big breeze capacity" as advertised and was soon discontinued from production.

Left: ***Emerson***
Type: Seagull
Blades: 9"
Date: 1934 – 35

Right: ***Emerson***
Type: Seagull
Blades: 10"
Date: 1934 – 35

In 1934 Emerson introduced two lines of new fans, the Seagull and Silver Swan. The Seagull was an inexpensive line of fans that Emerson hoped would tap into the lower priced market. This fan was designed to compete with dime store fans of the period. It was advertised as Emerson's masterpiece but that was far from the truth. The motor was very poorly made and had not much more than cardboard bearings. The Art Deco design was to give the motor a "floating in air" look. There was no switch on the fan and the blades were made of thin steel. A chrome seagull is located on top of the guard with two chrome wires coming out of the seagull towards the base. The base is cast-iron with a paper decal on the front stating the name of the fan. This was truly the worst fan ever made by Emerson. It did not compete well with the lower price fans of the period and made the company little money. Production was soon dropped for more profitable fans.

Grade I – $90.00
Grade II – $68.00
Grade III – $45.00
Grade IV – $32.00
Grade V – $18.00
Grade VI – $5.00

The larger Seagull fan was better designed than the 9". It had a cast-iron base with a single speed toggle switch. The guard incorporated a chrome seagull on the top of the guard with two chrome wires coming out of the seagull down toward the base giving it a very Deco look. The side of the guard is made of sheet metal with spot welded wires. The motor is better constructed than the 9" version, but not by much. The oscillator was a very cheaply made device that worked, but was far from the excellent oscillators on Emerson's better grade fans. The traditional scallop Parker blade were incorporated into the design to give it that distinct Emerson look. Like the 9" Seagull fans, the 10" models did not compete well with the lower priced dime store fans of the period and were soon dropped from production.

Grade I – $150.00
Grade II – $113.00
Grade III – $75.00
Grade IV – $53.00
Grade V – $30.00
Grade VI – $8.00

Left: ***Emerson***
Type: 5250-B Silver Swan
Blades: 10"
Date: 1934 – 1937

Center: ***Emerson***
Type: 7250 Silver Swan Imperial
Date: 1937 – 1940

Right: ***Emerson***
Type: 5250-F Silver Swan
Blades: 10"
Date: 1940

In 1934 Emerson introduced the newly streamlined Silver Swan to boost sagging fan sales. The Silver Swan fans were produced from 1934 until 1942 when World War II began and aluminum became scarce. Following General Electric's lead, Emerson introduced an upscale quiet blade fan with lots of modern styling. The fan was a huge success. Several variations of this fan were made as shown above. The first example is a prototype of the first Silver Swan. The body of the fan is made of formed sheet steel attached to a steel wire frame. The blades are large and overlapping, finished in highly polished aluminum. The base is cast-iron and streamlined. The back of the fan (not seen) is covered by a chrome bullet. The motor has been newly designed and is almost completely hidden from view. An oscillator wheel is located at the base of the back of the motor held on by an extended shaft. The dark brown finish was an attempt to give the fan a less industrial look. The traditional slide switch has been replaced by a more modern looking chrome-plated toggle switch. All of the 10" Silver Swans were one speed. These were very expensive fans when manufactured and were targeted to more upscale markets.

Grade I – $300.00
Grade II – $225.00
Grade III – $150.00
Grade IV – $105.00
Grade V – $60.00
Grade VI – $15.00

The Imperial Silver Swan was a modification of the 1934 version with an all chrome body. The aluminum blades are painted flat black to contrast with the highly polished chrome body. The base design is more elongated and ribbed than the first version. The toggle switch has been moved from the side of the fan to the front. These were very expensive fans and only made in the 10". They will oscillate and have one speed.

Grade I – $400.00
Grade II – $300.00
Grade III – $200.00
Grade IV – $140.00
Grade V – $80.00
Grade VI – $20.00

This example is a redesign of the 1934 Silver Swan with a new highly polish chrome front screen attached to a round sheet-steel body. The base is cast metal and ribbed like the Imperial. In front of the base is a small stamped metal tag. Most of the Silver Swans had paper decals for badges a legacy of the Seagull period.

Grade I – $300.00
Grade II – $225.00
Grade III – $150.00
Grade IV – $105.00
Grade V – $60.00
Grade VI – $15.00

Left: ***Emerson***
Type: 5460 Silver Swan
Blades: 12"
Date: 1937 – 40

Right: ***Emerson***
Type: 4660 Silver Swan
Blades: 12"
Date: 1934

The 12" Silver Swan is massive and very impressive. The larger Silver Swans have three speeds and are very efficient wind machines. This example has a highly polished chrome screen attached to a round sheet-steel body. The blades are overlapping and finished in a highly polished aluminum. The base is made of cast metal and deco in design. The base contains a three-speed rotary switch for easy use.

Grade I – $500.00
Grade II – $375.00
Grade III – $250.00
Grade IV – $175.00
Grade V – $100.00
Grade VI – $35.00

This example has the original 1934 guard design with a cast ribbed base. The motor on the Silver Swan was not a single bearing as in Emerson's better grade fans, but two bearing with oil holes located at the top of the motor.

Grade I – $500.00
Grade II – $375.00
Grade III – $250.00
Grade IV – $175.00
Grade V – $100.00
Grade VI – $35.00

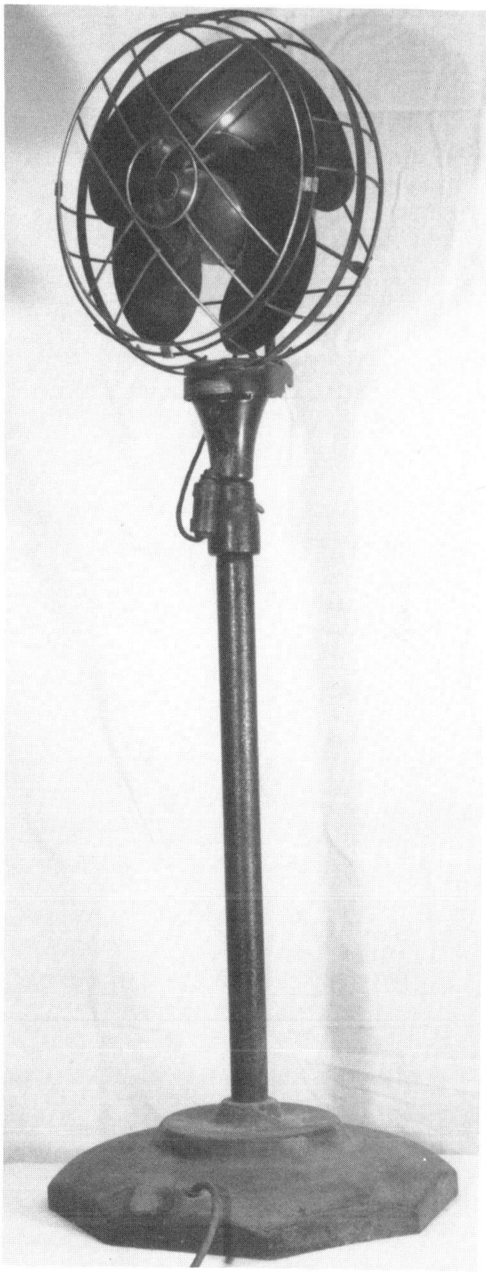

Emerson

Type: Silver Swan Pedestal
Blades: 12"
Date: c.1937

Emerson manufactured the Silver Swan in a pedestal fan as well as desk model. These fans were primarily marketed for commercial use, but were also purchased for residential use. This fan is massive by any account and was made of the finest materials. The base is cast-iron and contains a three-speed slide switch. The guard rests on a spring mounted frame that allows the fan to be tilted forward or backward. The motor is attached to the base by the guard with three heavy struts. A chrome screen is located at the front of the guard and is the only chrome plated part of the fan. The base has a wrinkle finish and is a dark brown mahogany color. The blades are painted black like on the Imperial Silver Swan. Emerson also manufactured Silver Swan pedestal fan in 10" model. The Silver Swan pedestal is one of the most collectible fans manufactured by Emerson.

Grade I –	$700.00	Grade IV –	$245.00
Grade II –	$525.00	Grade V –	$140.00
Grade III –	$350.00	Grade VI –	$35.00

Special Features that Make EMERSON the Long Life Fans

Exclusive, Emerson Bearing Design—Force-feed Lubrication System—Finger-tip Oscillation Control—Latch Handle for Centering Breeze—Durable Finish.

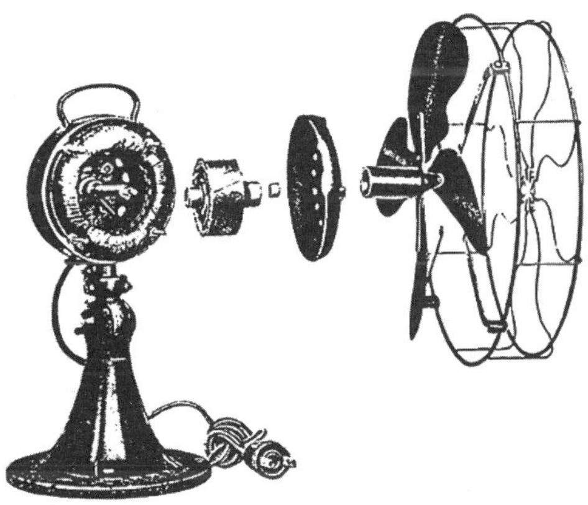

By showing the assembly of armature and blade on Emerson Standard 12 and 16-inch fans, you can give a practical demonstration of the extremely fine construction and simplicity of design.

The armature rotates on a stationary ½-inch diameter, case hardened, hollow steel shaft pressed into the motor frame. The armature core is porous cast iron, and this cast iron core and case hardened steel shaft make an ideal combination of bearing materials. The cast iron core absorbs oil and the bearing surfaces of both core and shaft acquire a fine glazed appearance like plated metal, which makes practically an everlasting bearing. Each armature core is hand reamed to fit the hollow shaft with which it is used.

This unique oil-tight, dust-proof bearing has been an exclusive feature on Emerson fans for over 30 years.

Force-Feed Lubrication

The lubricant in the fan hub is carried through the hollow shaft by the coil spring attached to the floating worm shaft. When the lubricant reaches the back end of the shaft, it is forced out through the slots in the hollow shaft to the bearing surfaces. The rotor core surface has a spiral groove which carries the lubricant forward, lubricating the entire length of the shaft as it travels to the fan hub. When additional lubricant is added through the oil well, it flows down through the hollow shaft to be picked up and put in circulation the same as the original lubricant.

Finger-Tip Oscillation Control

FIG. A

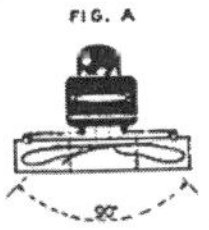

FIG. B

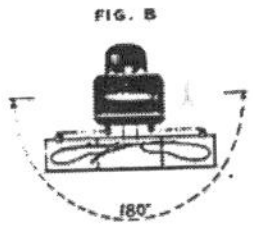

For ease of control—simplicity of adjustment and smooth operation, the exclusive Emerson oscillation adjustment is one of the outstanding achievements in the Emerson design. Now, finger-tip oscillation adjustment is made available on all Emerson oscillators, from the small 8-inch to the large 16-inch fans. Adjustment from non-oscillating position to any range of oscillation, up to 90°, is obtained by simply turning the rim of the oscillating adjusting case.

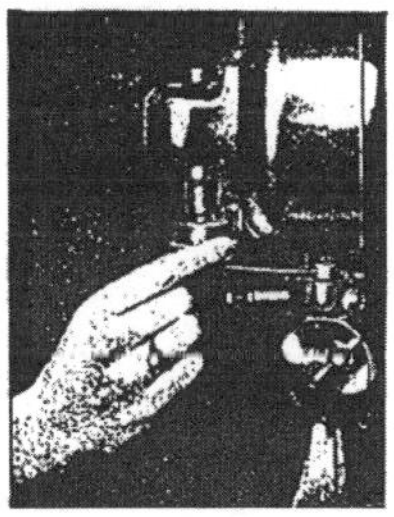

The exclusive "patented" adjusting collar on Emerson 12-inch and 16-inch fans is another important feature—shown directly below oscillating link in illustration at right. The breeze can be directed to any desired point, in a wide arc, without moving the base, and without the use of tools. This is particularly advantageous when the fan is mounted on the wall.

Figure A, above, illustrates the 90° maximum oscillation arc, and Figure B the complete range of oscillation possible, giving a total of 180°, from a fixed position through the five stops in the adjusting collar.

An illustration from a 1937 fan catalog noting two special features of Emerson fans; the forced-fed lubrication system and the finger-tip oscillation control. Emerson advertised that their fans would never spray the room with excess oil after being lubricated. This promise brought Emerson many loyal customers. The oscillating mechanism was easy to use and adjustable and was far superior to GE fans of the period.

Left: ***Emerson***
Type: 77648-AK
Blades: 16"
Date: 1937

In 1936 Emerson began experimenting with large overlapping blades as a way to reduce the noise made by small shallow blades. The first attempt was the series of fans shown above. These fans were made from 1936 to about 1939 and met with some success. Fan catalogs from 1936 suggest that this fan was a standard factory product. The blades are the only thing changed in this new design from the typical thirties desk fan. The guard has the traditional S wire with a Pyramid badge. The consumer could order a capacitor on this fan to reduce electrical consumption. The one pictured above has this option. Note how the slide switch is slightly raised above the rim of the base, the fan next to it does not have a capacitor and the switch sits lower. A capacitor could lower energy consumption by 40%, a major consideration if you were purchasing 100 fans to cool your factory, college, or hospital. This feature was only offered on this model of fans when first produced in the mid-1930s.

Grade I – $100.00
Grade II – $75.00
Grade III – $50.00
Grade IV – $35.00
Grade V – $20.00
Grade VI – $5.00

Center: ***Emerson***
Type: 77646
Blades: 12"
Date: 1936

This fan was made for commercial use. The newly designed overlapping blades were touted as being quieter and more efficient. Unfortunately, the shallow pitched overlapping blades were not much quieter than their scallop Parker blade counterparts. It was not until 1939 that Emerson perfected the large overlapping blade fan. One drawback to these larger blades is that they tend to become out of balance making the fan vibrate when in operation.

Grade I – $70.00
Grade II – $53.00
Grade III – $35.00
Grade IV – $25.00
Grade V – $14.00
Grade VI – $3.00

Right: ***Emerson***
Type: 6250
Blades: 10"
Date: 1937

This was an attempt to take a very popular 10" home use fan and put overlapping blades on it. This met with minimal success. Like the junior fans before it, this fan had all the information on the front badge. The six S wire guard was held to the motor by four bolts and had a single speed slide switch in the base. The motor housing and base were cast-iron and finished in gloss black. Many of these fans had vibration problems due to the blades being off balance.

Grade I – $40.00
Grade II – $23.00
Grade III – $20.00
Grade IV – $14.00
Grade V – $8.00
Grade VI – $2.00

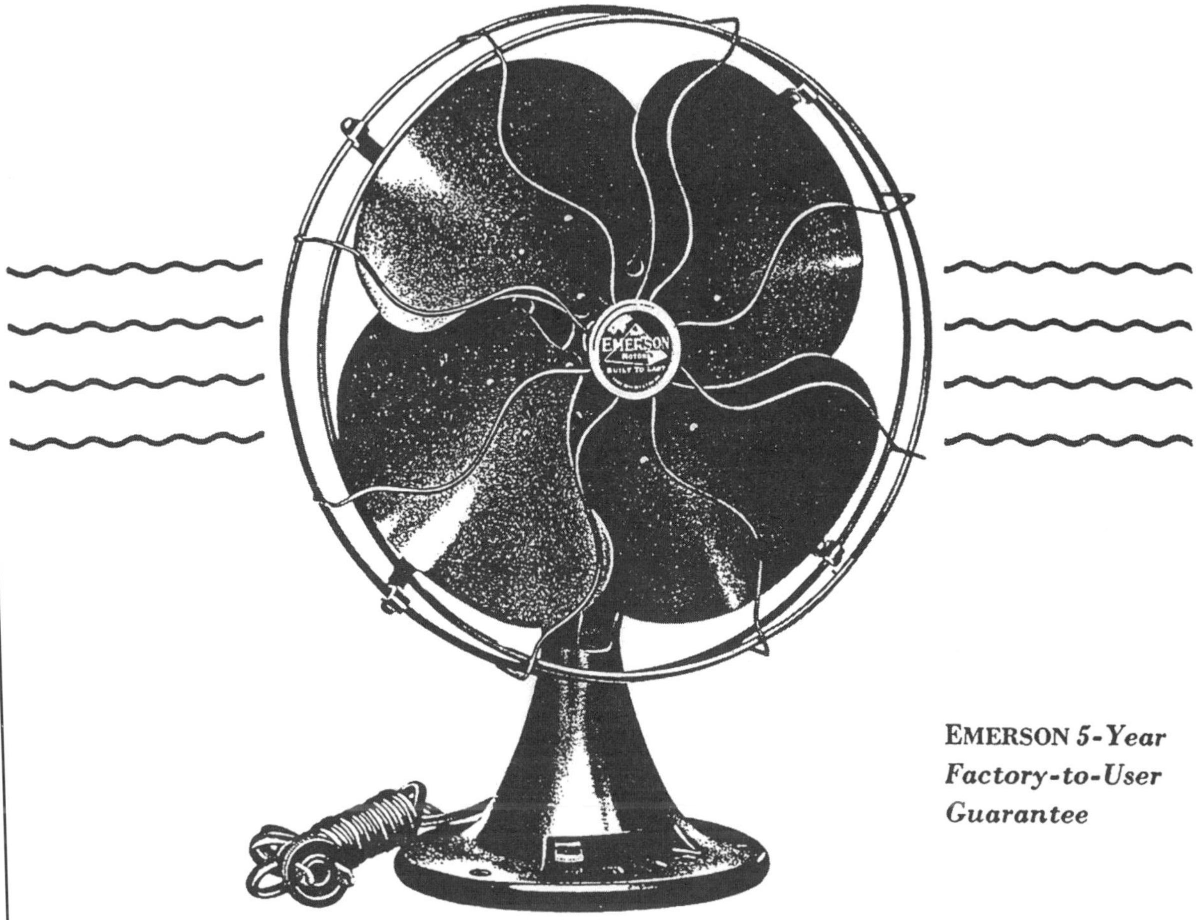

EMERSON 10, 12 and 16-inch Fans with Patented Overlapping Blades

SIZE	STYLE	VOLTS	CURRENT	SPEEDS AND WATTS	*AIR DELIVERY	WEIGHT NET	WEIGHT PKD.	TYPE No.	CODE WORD	LIST PRICE EACH
10″	Osc.	110	60 cy.	1550 rpm 40 watts Single-speed	610 CFM	12	17	6250	SECOL	$14.45
10″	Osc.	110	60 cy.	1550 rpm Single-speed 40 watts With adj. floor stand	610 CFM	21	29	6250-AA	SOGOR	19.95
12″	Osc.	110	60 cy.	1550 rpm 1340 rpm 1200 rpm 45 watts 42 watts 39 watts	1080 CFM	22	32	77646-AK	SOPEX	28.00
16″	Osc.	110	60 cy.	1500 rpm 1380 rpm 1150 rpm 60 watts 55 watts 50 watts	1550 CFM	24	38	77648-AK	SOHAN	35.00

* CFM—Cubic feet of air per minute on high speed.

Regularly carried in stock at St. Louis, New York, Chicago and by conveniently located wholesalers.

Fans will operate satisfactorily on voltage 10% above or below rated voltage. 50-cycle speed approximately 15% lower.

Special finish fans furnished to special order—see Price List.

No fans made for voltages higher than 250 volts. Performance data subject to variation of 10% plus or minus (all data subject to change).

All fans packed one in a box. (Type 6250-AA packed completely assembled.)

An illustration from a 1937 fan catalog introducing Emerson's newly designed overlapping blade fans. The only difference between this fan and the standard scalloped blade desk fan was the large overlapping blades. By the late thirties all major fan manufactures were producing some form of overlapping blade design. The previous year Emerson introduced this fan with a capacitor in the base to reduce electricity consumption. The fan pictured does not have this option. These fans reduced noise but were soon replaced by deeper pitched slower fans that were much quieter.

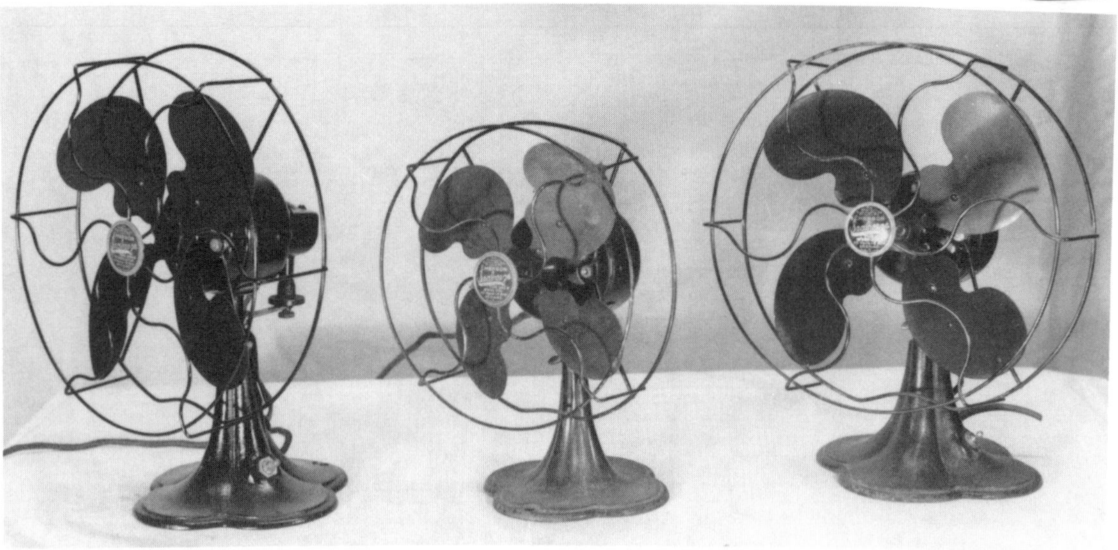

Left: ***Emerson***
Type: 3250-B Seabreeze
Blades: 10"
Date: 1936

In 1936 Emerson came out with a newly designed group of fans marketed under the name Seabreeze. After the Seagull fiasco Emerson learned that quality does count to the consumer. These fans were targeted to the lower price market than their Emerson grade fans. Emerson lowered cost by using a newly designed motor that had a stamped steel motor housing held together by two steel machine screws. These screws held the field coils, motor housing, and guard together saving cost in manufacturing. The base was made of cast-iron in a lily configuration. The traditional slide switch was replaced by a newer, more modern looking toggle switch. The guard was made of high quality steel wire in an S configuration. Like on the 10" Seagull, the scalloped Parker blade was retained. These fans were made in 8", 10", 12", and 16" sizes, and were very successful. The pictured example is an oscillator, but you could purchase any Seabreeze as a non-oscillator. The oscillator was different from their top of the line Emersons because you had to unscrew a set screw underneath the oscillator wheel to engage the oscillator. This could be very messy if oil or grease were to have dripped down from the gear box, however this did not disrupt the efficiency of the oscillator which worked very well.

Grade I – $60.00
Grade II – $45.00
Grade III – $30.00
Grade IV – $21.00
Grade V – $12.00
Grade VI – $3.00

Center: ***Emerson***
Type: 3140-B Seabreeze
Blades: 8"
Date: 1936

Smallest of the Seabreeze line and was only offered in a non-oscillator. These small fans had a toggle switch on the back of the motor. Not since the Emerson Juniors of the early 1920s had Emerson made an 8" fan with scalloped Parker blades.

Grade I – $30.00
Grade II – $23.00
Grade III – $15.00
Grade IV – $10.00
Grade V – $6.00
Grade VI – $2.00

Right: ***Emerson***
Type: 3150-B Seabreeze
Blades: 10"
Date: 1936

An example of a non-oscillator, the 12" and 16" versions were two speeds, but all 10" were single speed.

Grade I – $40.00
Grade II – $23.00
Grade III – $20.00
Grade IV – $14.00
Grade V – $8.00
Grade VI – $2.00

Left: ***Emerson***
Type: 79646-AN
Blades: 12"
Date: 1939

Right: ***Emerson***
Type: 79648-AX
Blades: 16"
Date: 1941

As the end of the thirties approached Emerson introduced fans with newly designed blades and guard. This new guard was more modern looking than the traditional S wire had been and had a cross shape to it that was attached to the motor by the use of a spring type latch. This latch allowed the front of the guard to be removed easily. The new blade design was a configuration of larger overlapping blades, but not as deeply pitched. This blade was to be used until production of fans ceased in the middle 1960s. This guard is almost exclusively used with bullet back fans, except for this example. This example is also important because the blades are polished aluminum. Emerson very seldom used polished aluminum blades on their fans with the exception of the Silver Swans. The base is designed for a capacitor. From 1939 on almost all Emerson fan bases on 12" and 16" fans were designed for capacitors, whether they had them or not.

Grade I – $90.00
Grade II – $68.00
Grade III – $45.00
Grade IV – $32.00
Grade V – $18.00
Grade VI – $5.00

This fan became a prototype of many Emerson fans made between 1940 and 1970. This fan incorporates the cross guard with a blue and silver badge. Newly designed Parker blades are attached to the motor by a stamped steel hub. A bullet has been added to the rear of the fan to hide the oscillator and give the fan a sleeker, more streamline look. The finish is gloss black with a felt base.

Grade I – $70.00
Grade II – $53.00
Grade III – $35.00
Grade IV – $25.00
Grade V – $14.00
Grade VI – $3.00

Emerson
Type: 79646-AP
Blades: 12"
Date: 1942

This fan was made for the government during World War II and followed their specifications. It seems that according to the federal government all fans had two bearings, so Emerson made a fan with two bearings. The oiliers are set at an angle which is an unusual feature. The Parker blades used on this fan are what were commonly used on direct current fans of the period. The base is very ugly and does not incorporate the nice wall mount ball found on top-of-the-line Emerson fans of the period. The base is fitted for a capacitor and it seems that this was a standard feature of this fan. The guard is painted black and there is no bullet hiding the oscillating mechanism. Most examples are painted in a wrinkle finish, but I have seen some in gloss black. What makes these fans unusual is that they look like a typical hollow shaft single bearing Emerson fan but the oilers and ugly base illustrate that this is not the case.

Grade I	– $60.00	Grade IV	– $21.00
Grade II	– $45.00	Grade V	– $12.00
Grade III	– $30.00	Grade VI	– $3.00

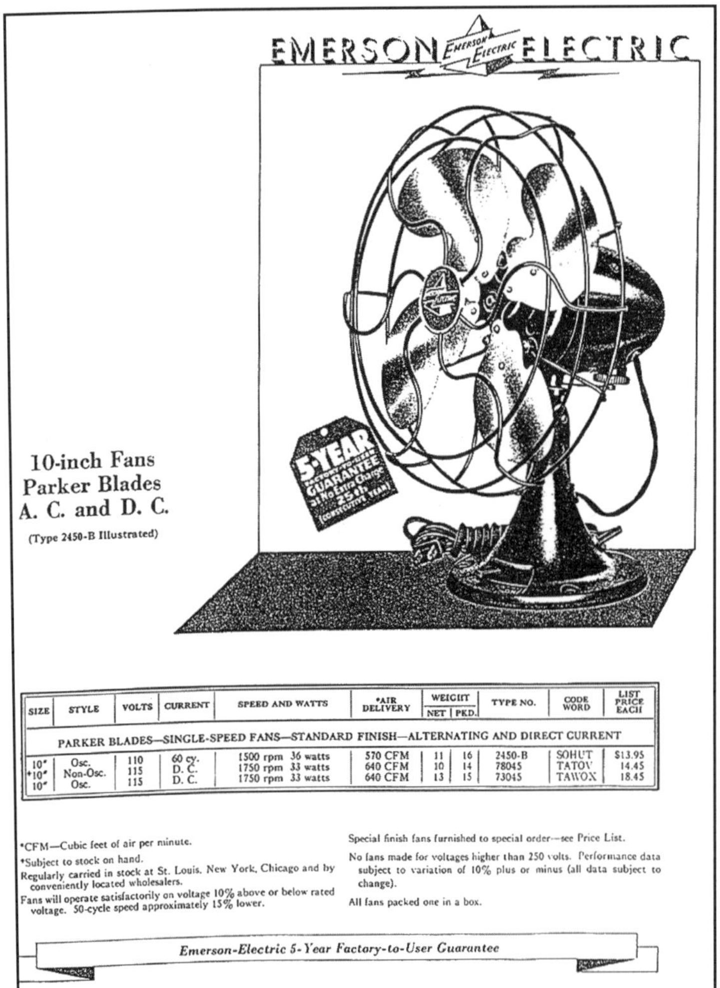

EMERSON ELECTRIC

10-inch Fans
Parker Blades
A. C. and D. C.
(Type 2450-B Illustrated)

SIZE	STYLE	VOLTS	CURRENT	SPEED AND WATTS	*AIR DELIVERY	WEIGHT NET	WEIGHT PKD.	TYPE NO.	CODE WORD	LIST PRICE EACH
PARKER BLADES—SINGLE-SPEED FANS—STANDARD FINISH—ALTERNATING AND DIRECT CURRENT										
10"	Osc.	110	60 cy.	1500 rpm 36 watts	570 CFM	11	16	2450-B	SOHUT	$13.95
†10"	Non-Osc.	115	D. C.	1750 rpm 33 watts	640 CFM	10	14	78045	TATOV	14.45
10"	Osc.	115	D. C.	1750 rpm 33 watts	640 CFM	13	15	73045	TAWOX	18.45

*CFM—Cubic feet of air per minute.
†Subject to stock on hand.
Regularly carried in stock at St. Louis, New York, Chicago and by conveniently located wholesalers.
Fans will operate satisfactorily on voltage 10% above or below rated voltage. 50-cycle speed approximately 15% lower.

Special finish fans furnished to special order—see Price List.
No fans made for voltages higher than 250 volts. Performance data subject to variation of 10% plus or minus (all data subject to change).
All fans packed one in a box.

Emerson-Electric 5-Year Factory-to-User Guarantee

An illustration from a 1939 fan catalog showing a 10" desk fan. This fan became the prototype for small desk fan design for a decade to come. Emerson streamlined the guard and added a bullet back to their popular 10" desk fan to come up with this configuration. Emerson manufactured some form of this design until 1950. The later examples have a toggle switch and wrinkle finish but are essentially the same fan as the earlier models.

Left: ***Emerson***
Type: 79646-AT
Blades: 12"
Date: 1941

Center: ***Emerson***
Type: 79646-AX
Blades: 12"
Date: 1941

Right: ***Emerson***
Type: 77646-AS
Blades: 12"
Date: 1939

During the 1940s Emerson made many fans with a wrinkle finish and this is an example. The wrinkle finish added texture and style to the fan, but had several drawbacks. The finish would attract dirt and hold oil if dropped on the fan. After the finish became soiled it was almost impossible to clean the fan without ruining the paint. Another drawback is that the paint itself was not as durable as the high gloss finish had been. A minor bump could gouge a hole in the finish making the fan unsightly. The blades are finished in a flat black. The guard badge is blue and silver. Note that the base is designed for a capacitor. The handle on this fan is large and roomy making for easy transport.

Grade I – $70.00
Grade II – $53.00
Grade III – $35.00
Grade IV – $25.00
Grade V – $14.00
Grade VI – $3.00

Another example of a pre-World War II fan made for industrial use. This has a high gloss black finish with Parker blades and a cross guard. These fans were very popular and were manufactured in large quantities.

Grade I – $60.00
Grade II – $45.00
Grade III – $30.00
Grade IV – $21.00
Grade V – $12.00
Grade VI – $3.00

This fan was most likely the very best overlapping blade desk fan made by Emerson. It was quiet, efficient, and beautiful. The large deeply pitched overlapping blades made it whisper quiet. It had a brown metallic finish with matching guard. The badge was gold and brown with a bullet back. These fans could be used for both commercial and residential use. The 16" fans are massive with their huge overlapping blades. These fans were also made after World War II in a black finish. Nothing on the market could compete with this fan in 1940.

Grade I – $90.00
Grade II – $68.00
Grade III – $45.00
Grade IV – $32.00
Grade V – $18.00
Grade VI – $5.00

EMERSON ELECTRIC

10, 12 and 16-inch Fans

Patented Overlapping Blades

A. C. and D. C.

(Type 77646-AG Illustrated)

SIZE	STYLE	VOLTS	CURRENT	SPEEDS AND WATTS	*AIR DELIVERY	WEIGHT NET	WEIGHT PKD.	TYPE NO.	CODE WORD	LIST PRICE EACH
OVERLAPPING BLADES—STANDARD FINISH—ALTERNATING CURRENT										
10"	Osc.	110	60 cy.	1550 rpm 40 watts Single-speed	610 CFM	12	16	6250-B	SOFUR	$15.45
10"	Osc.	110	60 cy.	1550 rpm 40 watts {Single-speed with adj. floor stand}	610 CFM	21	30	6250-AA	SOGOR	19.95
12"	Osc.	110	60 cy.	1550 rpm 45 watts; 1340 rpm 42 watts; 1200 rpm 39 watts	1080 CFM	22	33	77646-AG	SOCUN	29.95
12"	Osc.	110	60 cy.	{1550 rpm 45 watts; 1340 rpm 42 watts; 1200 rpm 39 watts; With adj. floor stand.}	1080 CFM	43	55	77646-AD	SELAR	36.45
16"	Osc.	110	60 cy.	1060 rpm 61 watts; 890 rpm 51 watts; 750 rpm 45 watts	1700 CFM	24	39	77648-AG	SOFIN	37.50
16"	Osc.	110	60 cy.	{1060 rpm 61 watts; 890 rpm 51 watts; 750 rpm 45 watts; With adj. floor stand.}	1700 CFM	48	65	77648-AH	SOWED	44.00
OVERLAPPING BLADES—CHROMIUM PLATED FINISH—ALTERNATING CURRENT										
12"	Osc.	110	60 cy.	1550 rpm 45 watts; 1340 rpm 42 watts; 1200 rpm 39 watts	1080 CFM	22	33	77646-AK	SOBAH	38.45
16"	Osc.	110	60 cy.	1550 rpm 60 watts; 1380 rpm 55 watts; 1150 rpm 50 watts	1550 CFM	24	39	77648-AK	SODIM	46.00
OVERLAPPING BLADES—STANDARD FINISH—DIRECT CURRENT										
10"	Osc.	115	D. C.	1750 rpm 33 watts; 1300 rpm 23 watts; 900 rpm 20 watts	675 CFM	13	17	77045	TAPEN	21.45
12"	Osc.	115	D. C.	1650 rpm 60 watts; 1300 rpm 45 watts; 900 rpm 35 watts	1150 CFM	25	38	77046-AA	TAFUH	33.45

*CFM—Cubic feet of air per minute on high speed.

Regularly carried in stock at St. Louis, New York, Chicago and by conveniently located wholesalers.

Fans will operate satisfactorily on voltage 10% above or below rated voltage. 50-cycle speed approximately 15% lower.

Special finish fans furnished to special order—see Price List.

No fans made for voltages higher than 250 volts. Performance data subject to variation of 10% plus or minus (all data subject to change).

All fans packed one in a box. (Types 6250-AA, 77646-AD and 77648-AH packed completely assembled.)

Emerson-Electric 5-Year Factory-to-User Guarantee

An illustration from a 1939 fan catalog introducing Emerson's newly designed overlapping blade fan. This fan was Emerson's finest overlapping blade fan and incorporated many new features. The guard had been completely redesigned and is a cross type. The slide switch has been placed higher up on the base to allow for a capacitor. This base became a standard feature on all 12" and 16" fans after this period. These fans were finished in a dark metallic brown and were very quiet fans.

Left: ***Emerson***
Type: 6250-D
Blades: 10"
Date: 1939

Right: ***Emerson***
Type: 6250-G
Blades: 10"
Date: 1939

As Emerson approached its fiftieth year of business they came out with a line of fans commonly known as the fifty-year or Jubilee fan. These are two examples. They were manufactured from about 1938 to 1942 and were produced in great quantities. These fans incorporated overlapping brass blades that were sometimes plated in a copper finish. The motor and base are made of cast-iron with a stamped steel bullet covering the back of the motor. The base incorporates a single speed toggle switch which replaced the traditional slide switch formerly used on this type of small fan. The cross-type guard is connected to the motor by four small rounded machine nuts. The first fan pictured is finished in a metallic brown color which is the most commonly used finish.

Grade I – $70.00
Grade II – $53.00
Grade III – $35.00
Grade IV – $25.00
Grade V – $14.00
Grade VI – $3.00

This is an example of the fifty-year fan in an ivory finish. The badge is painted the same color as the rest of the fan, noting a factory paint job. The front badge and motor badge are trimmed in highly polished brass. The lighter color finish is unusual and there were not as many of these fans manufactured. Unfortunately the lighter color finish did not hold up as well as the brown metallic and most of these examples are chipped and discolored.

Grade I – $130.00
Grade II – $98.00
Grade III – $65.00
Grade IV – $45.00
Grade V – $26.00
Grade VI – $7.00

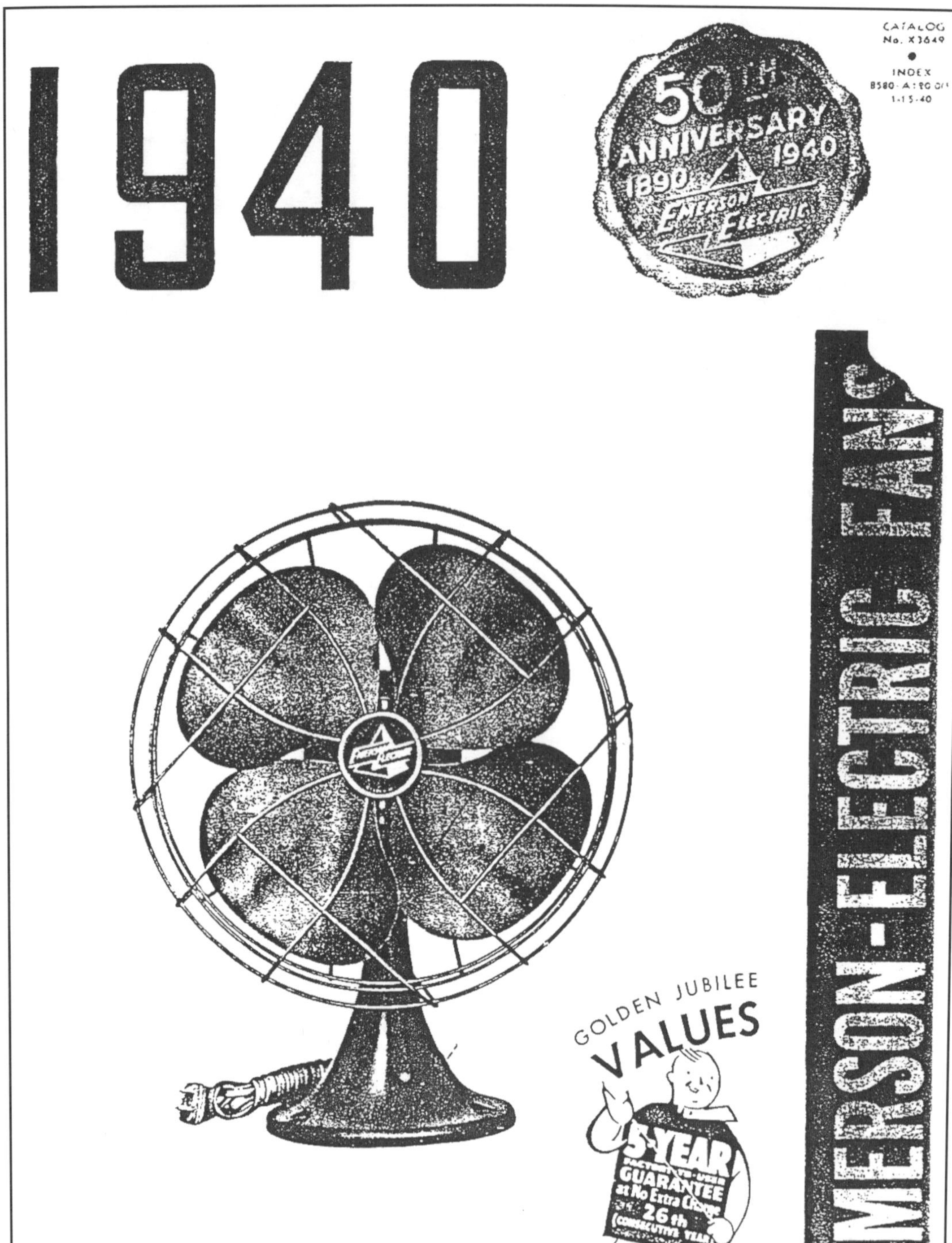

An illustration of a front cover of a 1940 fan catalog. This 10" fifty-year or Jubilee fan has a newly designed cross guard, overlapping brass blades, and a streamlined bullet back. The slide switch has been replaced with a toggle and these fans were manufactured in great numbers. Note the fifty-year seal located at the top right-hand corner of the catalog.

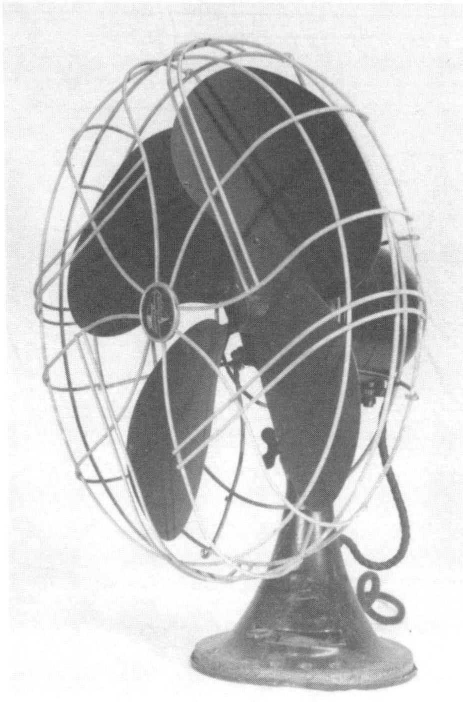

Emerson
Type: 77648-AL
Blades: 16"
Date: c.1940

This fan incorporates newly designed overlapping blades and a cross guard. The base and motor housing are made of cast-iron. The base has a capacitor for greater efficiency. The large blades make for a very dramatic large fan. This fan would have been marketed for industrial or commercial use. It is finished in high gloss back and will oscillate.

Grade I – $90.00
Grade II – $68.00
Grade III – $45.00
Grade IV – $32.00
Grade V – $18.00
Grade VI – $5.00

Right: ***Emerson***
Type: 2450-C
Blades: 10"
Date: 1949

Left: ***Emerson***
Type: 77645-AU
Blades: 12"
Date: 1940

This is a classic example of a special finish fan made by Emerson. This fan is finished in light yellow with matching badge. This fan was found with a group of three others in St. Louis, Missouri, and was in mint condition for a light colored fan. The guard badge is trimmed in gold, as well as the motor badge. It seems that these fans were never used and makes a rare addition to any collection.

Grade I – $150.00
Grade II – $113.00
Grade III – $75.00
Grade IV – $53.00
Grade V – $30.00
Grade VI – $8.00

Following World War II Emerson came out with a line of fans designed for residential use to tap into the post-war boom for manufactured goods. This fan incorporates a newly streamlined designed guard that mixes the traditional S wire style used on fans after the war. This was a change from pre-war era fans that had the badges that matched the finish of the fan. The blades are no longer overlapping, but modified 16" Parker blades. These blades are very attractive, but almost look like blades made by General Electric in the 1920s. This fan was the top of the line model because it had two speeds, unusual for a 10" fan. It has a cream colored wrinkle finish.

Grade I – $40.00
Grade II – $23.00
Grade III – $20.00
Grade IV – $14.00
Grade V – $8.00
Grade VI – $2.00

Left: ***Emerson***
Type: 2250-D
Blades: 10"
Date: 1938

This fan incorporates a newly designed blade and guard configuration. The blades are very similar to the new 12" and 16" Parker blades so widely used after 1940. The guard is a streamlined S wire with a black and silver center badge. There is no motor badge, all information is found on the center badge. The base and motor housing are cast-iron. There is no bullet on the back of the fan, making this a very unusual example.

Grade I – $40.00
Grade II – $23.00
Grade III – $20.00
Grade IV – $14.00
Grade V – $8.00
Grade VI – $2.00

Right: ***Emerson***
Type: 2450-D
Blades: 10"
Date: 1949

An example of a mass-produced post-World War II fan. This fan still incorporates the traditional slide switch, but is painted in a black wrinkle finish. The blades are painted in a flat black and the badge is black and silver.

Grade I – $40.00
Grade II – $23.00
Grade III – $20.00
Grade IV – $14.00
Grade V – $8.00
Grade VI – $2.00

The fans pictured represent a typical Emerson 10" fan mass produced for residential or office use. They were top of the line models manufactured in large quantities for public consumption. All of these fans have bullet backs and newly streamlined S guards.

Left: ***Emerson***
Type: 2450-B
Blades: 10"
Date: 1939

This fan has scalloped Parker blades painted silver. Some examples have highly polished chrome blades. The fan is finished in a high gloss black.

Grade I – $30.00
Grade II – $23.00
Grade III – $15.00
Grade IV – $10.00
Grade V – $6.00
Grade VI – $2.00

Center: ***Emerson***
Type: 2450-D
Blades: 10"
Date: 1949

Same fan as pictured on page 58.

Grade I – $30.00
Grade II – $23.00
Grade III – $15.00
Grade IV – $10.00
Grade V – $6.00
Grade VI – $2.00

Right: ***Emerson***
Type: 2450-G
Blades: 10"
Date: 1951

This fan is the same as the other two pictured, but the slide switch has been replaced by a toggle switch. The blades are finished in a flat black and the rest of the fan in a high gloss black. By 1950 this fan must have looked rather old fashion.

Grade I – $30.00
Grade II – $23.00
Grade III – $15.00
Grade IV – $10.00
Grade V – $6.00
Grade VI – $2.00

The Emerson Junior line was reintroduced in 1940 and gave the consumer a new modern designed fan. The blades, guard, and base have all been reconfigured to convey a very modern and stylish look. The blades have been rounded, but are not overlapping. The guard had almost a cubist look to it with a bright red center badge. The base was very similar to the Seabreeze line of fans that came out in 1936.

Left: ***Emerson***
Type: 2650-B
Blades: 12"
Date: 1940

A two-speed fan with a cast-iron lily base. The motor housing is stamped steel with a large bullet back. The oscillator is similar to that found on Seabreeze models of the late thirties. The color is dark brown and very industrial looking. This fan has a similar oscillator to that found on Seabreeze fans.

Grade I – $30.00
Grade IV – $10.00
Grade II – $23.00
Grade III – $15.00
Grade V – $6.00
Grade VI – $2.00

Center: ***Emerson***
Type: 2540-B
Blades: 8"
Date: 1940

Small fans were popular for bedrooms, bathrooms, and kitchens, and marketed for such. This fan has a cast-iron base and bullet back with one speed. There is no switch on the fan and it is a non-oscillator.

Grade I – $30.00
Grade II – $23.00
Grade III – $15.00
Grade IV – $10.00
Grade V – $6.00
Grade VI – $2.00

Right: ***Emerson***
Type: 2260-H
Blades: 12"
Date: 1940

These junior fans could be purchased in several colors and this example is metallic green. This is not very apparent by the black and white picture, but the effect is dramatic. These fans also came in metallic red, another stunning color. The colored junior fans are more collectible than the brown ones because they were not produced in such great quantities.

Grade I – $50.00
Grade II – $38.00
Grade III – $25.00
Grade IV – $18.00
Grade V – $10.00
Grade VI – $3.00

Left: ***Emerson***
Type: 77648-AN
Blades: 16"
Date: 1940

This large fan has impressive overlapping blades with a double wire cross guard. The pairs of wires bisecting the cross are doubled on 16" fans of this period. The fan is finished in a dark brown wrinkle finish. These fans were very efficient and quiet and could be purchased with a capacitor for further energy reduction.

Grade I – $90.00
Grade II – $68.00
Grade III – $45.00
Grade IV – $32.00
Grade V – $18.00
Grade VI – $5.00

Left: ***Emerson***
Type: 77045
Blades: 10"
Date: 1938

Emerson made direct current fans primarily for industrial and commercial use, and these are two later examples. Note that the newly designed overlapping blades and cross guard have been incorporated into the design. It seems that Emerson never put a bullet back on direct current fans, even though they were a standard feature after about 1940. Another thing that makes this fan unusual is the small size. Ten inch fans were primarily used for residential and not for commercial use.

Grade I – $300.00
Grade II – $225.00
Grade III – $150.00
Grade IV – $105.00
Grade V – $60.00
Grade VI – $15.00

Right: ***Emerson***
Type: 77048-AK
Blades: 16"
Date: 1940

This is a very late example of a direct current fan made for industrial use. It incorporates all the latest fan modifications for 1940 including large overlapping blades, cross guard, but no bullet back. The fan is finished in a high gloss black enamel. The base has a raised switch designed for a capacitor, but this would be of no use on a direct current fan. This fan is extremely efficient and blows a large amount of air.

Grade I – $400.00
Grade II – $300.00
Grade III – $200.00
Grade IV – $140.00
Grade V – $80.00
Grade VI – $20.00

Left: ***Emerson***
Type: 77648-SO
Blades: 16"
Date: 1956

This is an example of a standard mass-produced fan of the middle fifties. This large fan has a silver finish with a cross type guard. The blades are the new Parker style used on later Emerson fans. The slide switch is still incorporated in this design as well as the ball at the top of the base for wall mount installation. These fans were produced in great quantities.

Grade I – $60.00
Grade II – $45.00
Grade III – $30.00
Grade IV – $21.00
Grade V – $12.00
Grade VI – $3.00

Right: ***Emerson***
Type: 79648-SD
Blades: 16"
Date: 1955

Emerson continued to manufacture fans with large overlapping blades well into the fifties for industrial and commercial use. This is a very late example. The fan is finished in a metallic gloss dark brown. This fan also incorporates the wall mount feature at the top of the base and is one of the last fans to do so.

Grade I – $50.00
Grade II – $38.00
Grade III – $25.00
Grade IV – $18.00
Grade V – $10.00
Grade VI – $3.00

Emerson
Type: 95643-B
Blades: 5"
Date: 1955

This is the only rubber bladed fan manufactured by Emerson. The base of the fan is actually a stamped steel ashtray modified to be used as a fan base. This fan is very quiet when operating, but was rather out of date when manufactured seeing that rubber bladed fans had reached their peak right before World War II. These were not very popular fans and were manufactured for a short time.

Grade I – $100.00
Grade II – $75.00
Grade III – $50.00
Grade IV – $35.00
Grade V – $20.00
Grade VI – $5.00

By the late 1950s Emerson was getting out of the desk fan business. Air conditioning made desk fans obsolete according to the prevailing thought of the time. New factory buildings didn't need hundreds of large 12" and 16" fans as they once had and the golden age of desk fans was drawing to a close. These two fans look very similar to the fans produced from 1940 but are not of the same high quality. There have been several major design changes in the switch and the base. The base has been modified and streamlined. The traditional wall mount ball feature has been removed and a much cheaper design has been used. The slide switch which had been used on Emerson's top-of-the-line fans since 1899 was dropped in favor of a rotary switch. There is no motor badge on this fan, only a small tag around the switch giving the type, model, and other information. One redeeming feature of this fan is that they would not throw oil when lubricated as guaranteed.

Left: ***Emerson***
Type: 79648-SM
Blades: 16"
Date: 1959

This fan incorporates new Parker blades and the cross guard. Note the large fancy badge, which was a common feature on the last fans manufactured by Emerson. The oscillator wheel on these last Emersons is often poorly constructed and will sometimes jam.

Grade I – $70.00
Grade II – $53.00
Grade III – $35.00
Grade IV – $25.00
Grade V – $14.00
Grade VI – $3.00

Right: ***Emerson***
Type: 77648-TE
Blades: 16"
Date: 1958

One of the last overlapping blade fans manufactured by Emerson having a large cross guard. The fan is finished in a dark metallic brown with a three-speed switch in the base. A very quiet fan and efficient air circulator.

Grade I – $60.00
Grade II – $45.00
Grade III – $30.00
Grade IV – $21.00
Grade V – $12.00
Grade VI – $3.00

Emerson made small desk fans for residential use until they stopped desk fan production. These are two examples of the last type of Emerson desk fan. The first example is an Emerson grade fan, the second a junior grade fan. Both have rounded blades, chrome-plated guards with a bullet back. The base is made of cast metal in a modern design. There are no electrical parts in the base. A two-speed push-in switch is located at the rear of the fan on the bullet. The only difference between these fans is that the guards are different.

Left: ***Emerson***
Type: 94646-D
Blades: 12"
Date: 1959

Grade I – $30.00
Grade II – $23.00
Grade III – $15.00
Grade IV – $10.00
Grade V – $6.00
Grade VI – $2.00

Right: ***Emerson***
Type: 2660-M Junior
Blades: 12"
Date: 1958

Grade I – $30.00
Grade II – $23.00
Grade III – $15.00
Grade IV – $10.00
Grade V – $6.00
Grade VI – $2.00

Emerson
Type: 94646-B
Blades: 12"
Date: 1955 – 1964

Production of these fans began in 1954 and lasted until about 1964. This example is one of the last fans manufactured by Emerson. The guard is made of concentric chrome-plated steel wires with a large plastic badge. This badge was the last one used on Emerson fans. The blades are round and similar to the junior fans first introduced in 1940. The motor is small and has a two-speed push-in switch. The modern style base is made of cast metal and has no electrical parts. This was the last desk fan made by Emerson, thus ending an era of desk fan manufacturing.

Grade I – $30.00
Grade II – $23.00
Grade III – $15.00
Grade IV – $10.00
Grade V – $6.00
Grade VI – $2.00

Emerson Numbers

Date	Type	Description and important details
1891	M1	12" Meston-Emerson, brush-com, tripod base, brass front & back motor plates
1891	M2	16" Meston-Emerson, brush-com, tripod base, brass front & back motor plates
1895	EI 1	12" 133 cy
1895	F 1	12" Emerson, tripod base, rotary swtich with manual start, 1st to use Parker blades
1895	F 2	16" Emerson, tripod base, rotary switch with manual start, also used Parker blades
1899	PI 241	12" Emerson, B/B, open ring guard, manual start in base, ribbed base
1899	PI 242	16" Emerson, B/B, open ring guard, manual start in base, ribbed base
1901	310	12" Emerson, B/B, Ribbed base, manual start, Parker blades
1901	311	12" Emerson, B/B, Ribbed base, manual start, Parker blades
1901	320	16" Emerson, B/B, Ribbed base, manual start, Parker blades
1901	321	16" Emerson, B/B, Ribbed base, manual start, Parker blades
1901	420	12" Emerson, B/B, Ribbed base, manual start, Parker blades
1901	421	16" Emerson, B/B, Ribbed base, manual start, Parker blades
1901	510	12" Emerson, B/B, Ribbed base, manual start, Parker blades
1901	511	12" Emerson, B/B, Ribbed base, manual start, Parker blades
1901	520	15" Emerson, B/B, Ribbed base, manual start, Parker blades
1901	512	15" Emerson, B/B, 133 cy, bracket swivel, Parker blades
1901	610	12" Emerson, B/B, desk model, Parker blades
1901	611	12" Emerson, B/B, desk model, Parker blades
1901	620	16" Emerson, B/B, desk model, Parker blades
1901	621	16" Emerson, B/B, bracket swivel, Parker blades
1902	710	12" Emerson B/B, bracket *first with Auto-Start, Parker Blades
1902	711	12" Emerson, B/B, 133 cy, Auto-Start, bracket with Parker Blades
1902	720	15" Emerson, B/B, 133 cy, Auto-Start, desk with Parker Blades
1902	721	15" Emerson, B/B, 133 cy, Auto-Start, bracket with Parker Blades
1902	810	12" Emerson, B/B, 133 cy, Auto-Start, desk model with Parker Blades
1902	811	12" Emerson, B/B, 133 cy, Auto-Start, bracket with Parker Blades
1902	820	15" Emerson, B/B, 133 cy, Auto-Start, desk model, Parker Blades
1902	821	15" Emerson, B/B, 133 cy, Auto-Start, bracket with Parker Blades
1902	910	12" Emerson, B/B, Auto-Start, desk model, Parker blades
1902	911	12" Emerson, B/B, Auto-Start, bracket with Parker blades
1902	920	16" Emerson, B/B, Auto-Start, desk model, Parker blades
1902	921	16" Emerson, B/B, Auto-Start, bracket with Parker blades
1902	1010	12" Emerson, B/B, Auto-Start, desk model, Parker blades
1902	1011	12" Emerson, B/B, Auto-Start, bracket with Parker blades
1902	1020	16" Emerson, B/B, Auto-Start, desk model, Parker blades
1902	1021	16" Emerson, B/B, Auto-Start, bracket with Parker blades
1903	1120	16" Emerson, B/B, Auto-Start, desk with swivel, Parker Blades
1903	1220	16" Emerson, B/B, Auto-Start, desk model, Parker Blades
1903	1221	16" Emerson, B/B, Auto-Start, bracket with Parker Blades
1903	1310	12" Emerson, B/B, Auto-Start, desk model, Parker Blades
1903	1311	12" Emerson, B/B, Auto-Start, bracket with Parker Blades
1903	1320	16" Emerson, B/B, Auto-Start, desk model, Parker Blades
1903	1321	16" Emerson, B/B, Auto-Start, bracket with Parker Blades

Date	Type	Description and important details
1904	5010	12" Trojan, B/B, 133 cy, non-auto start, desk model
1904	5011	12" Trojan, B/B, 133 cy, non-auto start, bracket
1904	5020	16" Trojan, B/B, 133 cy, non-auto start, desk mode
1904	5021	16" Trojan, B/B, 133 cy, non-auto start, bracket
1904	5110	12" Trojan, B/B, non-auto start, desk model
1904	5111	12" Trojan, B/B, non-auto start, bracket
1904	5120	16" Trojan, B/B, non-auto start, desk model
1904	5121	16" Trojan, B/B, non-auto start, bracket
1904	5210	12" Trojan, B/B, non-auto start desk model
1904	5211	12" Trojan, B/B, non-auto start bracket
1905	1015	12" Emerson, B/B, 3 speed, desk
1905	1016	12" Emerson, B/B, 3 speed, bracket
1905	1025	16" Emerson, B/B, 3 speed, desk
1905	1026	16" Emerson, B/B, 3 speed, bracket
1906	1115	12" Emerson, B/B, desk model, 3 speed, Parker Blades
1906	1215	16" Emerson, B/B, *first residential fan, six blade, 3 speed
1906	1510	12" Emerson, B/B, desk model, Parker Blades
1906	1520	16" Emerson, B/B, desk model, Parker Blades
1906	1610	12" Emerson, B/B, desk model, Parker Blades
1906	1620	16" Emerson, B/B, desk model, Parker Blades
1906	1710	12" Emerson, B/B, 133 cy, desk model, Parker Blades
1906	1720	15" Emerson, B/B, 133 cy, desk model, Parker Blades
1906	5310	12" Trojan, B/B, *Auto-Start divided pole moter, first with Auto-Start
1906	5320	16" Trojan, B/B, *Auto-Start divided pole motor
1906	5410	12" Trojan, B/B, non-auto start, solid pole motor
1906	5510	12" Trojan, B/B, 133 cy, non-auto start, solid pole motor
1906	5520	16" Trojan, B/B, 133 cy, non-auto start, solid pole motor
1908	1215	12" Emerson, B/B, *first 12" residential fan, six Parker blades
1908	1820	16" Emerson, B/B, desk model with Parker blades
1908	2010	12" Emerson, B/B, desk model with Parker blades
1908	2010a	12" Emerson, B/B, desk model with Parker blades
1908	2110	12" Emerson, B/B, 133 cy, desk model with Parker blades
1908	2120	16" Emerson, B/B, 133 cy, desk model with Parker blades
1908	5610	12" Trojan, B/B, Auto-Start, divided pole motor
1908	5710	12" Trojan, B/B, 133 cy, Auto-Start, divided pole motor
1908	5720	16" Trojan, B/B, 133 cy, Auto-Start, divided pole motor
1909	1500	8" Emerson, B/B, desk model with Parker blades, first 8" fan
1909	2210	12" Emerson, B/B, *first oscillator, lever type, Parker blades
1909	2220	16" Emerson, B/B, *first oscillator, lever type, Parker Blades
1909	2310	12" Emerson, B/B, 133 cy, desk oscillator, lever type, Parker blades
1909	2320	16" Emerson, B/B, 133 cy, desk oscillator, lever type, Parker blades

Date	Type	Description and important details
1910	11046	12" Emerson, B/B, auto-start, 3 speed desk with Parker Blades
1910	11066	12" Emerson, B/B, auto-start residential fan with six Parker Blades
1910	11644	8" Emerson, B/B, auto start desk model with Parker Blades
1910	11646	12" Emerson, B/B, auto-start desk model with Parker Blades
1910	11648	16" Emerson, B/B, auto-start desk model with Parker Blades
1910	11666	12" Emerson, B/B, auto-start residential fan with six Parker Blades
1910	11946	12" Emerson, B/B, 133 cy, desk model with Parker blades
1910	11948	16" Emerson, B/B, 133 cy, desk model with Parker blades
1910	12646	12" Emerson, B/B, ociallator, auto-start, lever type, Parker blades
1910	12648	16" Emerson, B/B, oscillator, auto-start, lever type, Parker Blades
1910	12666	12" Emerson, B/B, *first residential oscillator, lever type, 1 speed, Parker blades
1910	12946	12" Emerson, B/B, 133 cy, desk model with Parker blades
1910	12948	16" Emerson, B/B, 133 cy, desk model with Parker blades
1910	51646	12" Trojan, B/B, auto-start, desk model
1910	51648	16" Trojan, B/B, auto-start, desk model
1910	51946	12" Trojan, B/B, 133 cy, auto-start, desk model
1910	51948	16" Trojan, B/B, 133 cy, auto-start, desk model
1910	52646	12" Trojan, B/B, *first oscillator, lever type desk model
1910	52648	16" Trojan, B/B, *first oscillator, lever type desk model
1910	52946	12" Trojan, B/B, 133 cy, oscillator, lever type desk model
1910	52948	16" Trojan, B/B, 133 cy, oscillator, lever type desk model
1911	14666	12" Emerson, B/B, residential 3 speed desk model, non-osc
1911	14648	12" Emerson, B/B, 3 speed desk model with Parker Blades
1911	15644	8" Emerson Bedroom, B/B, 1 speed with cast bracket, first bedroom fan
1911	53644	8" Trojan, B/B, 3 speed oscillator, level type
1912	11044	8" Emerson, B/B, 3 speed, desk model, Parker blades
1912	14046	12" Emerson, B/B, 3 speed, desk model, Parker blades
1912	15044	8" Emerson, B/B, 1 speed, desk model, Parker blades
1912	14048	16" Emerson, B/B, 3 speed, desk model, Parker blades
1912	14066	12" Emerson, B/B, 3 speed desk model, Parker blades
1912	14646	12" Emerson, B/B, 3 speed desk model, Parker blades
1912	14648	16" Emerson, B/B, 3 speed desk model, Parker blades
1912	14946	12" Emerson, B/B, 133 cy, 3 speed oscillator, lever type
1912	14948	16" Emerson, B/B, 133 cy, 3 speed oscillator, lever type
1912	16646	12" Emerson, B/B, 3 speed, oscillator, lever type, Parker blades
1912	16648	16" Emerson, B/B, 3 speed oscillator, lever type, Parker blades
1912	16666	12" Emerson, B/B, 3 speed, residential oscillator, lever type, Parker blades
1912	16946	12" Emerson, B/B, 133 cy, 3 speed, oscillator, lever type, Parker blades
1912	16948	16" Emerson, B/B, 133 cy, 3 speed oscillator, lever type, Parker blades
1912	17666	16" Emerson, B/B, 3 speed, residential, Parker blades
1912	53646	12" Trojan, B/B, 3 speed, non-osc
1912	53648	16" Trojan, B/B, 3 speed non-osc
1912	54646	12" Trojan, B/B, 3 speed oscillator, lever type
1912	54648	16" Trojan, B/B, 3 speed oscillator, lever type

Date	Type	Description and important details
1913	14066	12" Emerson, B/B, 3 speed residential non-osc
1913	18046	12" Emerson, *first ventilating fan, 3 speed
1913	18048	16" Emerson, *first ventilating fan, 3 speed
1913	18646	12" Emerson, ventilating fan, divided pole induction motor
1913	18648	16" Emerson, ventilating fan, divided pole induction motor
1914	19044	8" Emerson, B/B, 3 speed desk model, no starting switch in motor, Parker blades
1914	19046	12" Emerson, B/B, 3 speed, no starting switch, oscillator, lever type, Parker blades
1914	19048	16" Emerson, B/B, 3 speed, no starting switch, oscillator, lever type, Parker blades
1914	19666	12" Emerson, B/B, 3 speed, no starting switch, oscillator, lever type, Parker blades
1914	21046	12" Emerson, B/B, 3 speed, new improved oscillator, Parker blades
1914	21048	16" Emerson, B/B, 3 speed, new improved oscillator, Parker blades
1914	21066	12" Emerson, B/B, 3 speed residential non-osc, Parker blades
1914	19246	12" Emerson, 25 cy, non-osc desk model, Parker blades
1914	19248	16" Emerson, 25 cy, non-osc desk model, Parker blades
1914	19645	9" Emerson, non-osc desk model, Parker blades
1914	19646	12" Emerson, non-osc desk model, Parker blades
1914	19648	16" Emerson, non-osc desk model, Parker blades
1914	19666	12" Emerson, residential desk model, Parker blades
1914	19946	12" Emerson, 133 cy, non-osc, desk model, Parker blades
1914	21246	12" Emerson, 25 cy, 3 speed, new improved oscillator, Parker blades
1914	21248	16" Emerson, 25 cy, 3 speed, new improved oscillator, Parker blades
1914	21646	12" Emerson, 3 speed, new improved oscillator, Parker blades
1914	21648	16" Emerson, 3 speed, new improved oscillator, Parker blades
1914	21666	12" Emerson, residential oscillator, lever type, Parker blades
1914	21946	12" Emerson, 133 cy oscillator, lever type, Parker blades
1915	19266	12" Emerson, 133 cy, non-osc, residential desk model, Parker blades
1915	21266	12" Emerson, 3 speed, residential oscillator, Parker blades
1915	21645	9" Emerson, 3 speed, oscillator, Parker blades
1915	21948	16" Emerson, 133 cy oscillator, Parker blades
1915	22646	12" Emerson ventilating fan
1915	22648	16" Emerson ventilating fan
1915	23045	9" Emerson, 3 speed, oscillator, Parker blades
1915	23046	12" Emerson, 3 speed, oscillator, Parker blades
1915	23048	16" Emerson, 3 speed, oscillator, Parker blades
1915	23666	12" Emerson, 3 speed, residential oscillator, Parker blades
1916	19644	8" Emerson, non-osc desk model, Parker Blades
1916	44A	8" Northwind, 2 speed, non osc *first Northwind series
1917	24246	16" Emerson, 25 cy, 3 speed, oscillator, Parker blades
1917	24266	12" Emerson, 25 cy, 3 speed, oscillator, Parker blades
1917	24645	9" Emerson, 3 speed, oscillator, Parker blades
1917	24646	12" Emerson, 3 speed, oscillator, Parker blades
1917	24648	16" Emerson, 3 speed, oscillator, Parker blades
1917	24666	12" Emerson, 3 speed, residential oscillator, Parker blades
1917	24668	16" Emerson, 3 speed, oscillator with Parker blades
1917	24946	12" Emerson, 133 cy, oscillator with Parker blades
1917	24948	16" Emerson, 133 cy, oscillator with Parker blades
1917	19045	9" Emerson, non-osc, 3 speed, Parker blades

Date	Type	Description and important details
1917	42045	12" Emerson, 3 speed, non-osc, desk model, Parker blades
1917	24046	16" Emerson, 3 speed, non-osc, desk model, Parker blades
1917	24666	12" Emerson, 3 speed, non-osc, residential, Parker blades
1917	44B	8" Northwind, 2 speed, non-osc
1917	450A	10" Northwind, 3 speed, oscillator, *first 10" model
1919	26646	12" Emerson, 3 speed, non-osc, desk model, Parker blades
1919	26648	16" Emerson, 3 speed, non-osc, desk model, Parker blades
1919	36046	12" Emerson, 3 speed, non-osc, desk model, Parker blades
1919	27046	16" Emerson, 3 speed, oscillator, Parker blades
1919	27048	16" Emerson, 3 speed, oscillator, Parker blades
1919	27248	16" Emerson, 25 cy, 3 speed, Parker blades
1919	27266	12" Emerson, 25 cy, 3 speed, residential non-osc, Parker blades
1919	27646	12" Emerson, 3 speed, oscillator, Parker blades
1919	27648	16" Emerson, 3 speed, oscillator, Parker blades
1919	27666	12" Emerson, 3 speed, residential oscillator, Parker blades
1919	44C	8" Northwind, 2 speed, non-oscillator
1919	450B	10" Northwind, 3 speed, oscillator
1920	22645	9" Emerson, 3 speed, oscillator, Parker blades
1920	26045	9" Emerson, 3 speed non-osc, Parker blades
1920	26048	16" Emerson, 3 speed, non-osc, desk model, Parker blades
1920	27045	9" Emerson, 3 speed, oscillator, desk model, Parker blades
1920	27645	9" Emerson, 3 speed, oscillator, desk model, Parker blades
1920	28645	9" Emerson, 3 speed, non-soc, desk model, Parker blades
1920	450C	10" Northwind, 3 speed, osc
1921	26266	12" Emerson, 25 cy, 3 speed, non-osc, residential
1922	28646	12" Emerson, 3 speed, non-osc, Parker blades
1922	28648	16" Emerson, 25 cy, 3 speed, ocillator, Parker blades
1922	29645	9" Emerson, 3 speed, oscillator, Parker blades
1922	29646	12" Emerson, 3 speed, oscillator, Parker blades
1922	29648	16" Emerson, 3 speed, oscillator, Parker blades
1922	29666	12" Emerson, 3 speed, residential oscillator, Parker blades
1922	29668	16" Emerson, 3 speed, non-osc, Parker blades
1922	30648	16" Emerson ventilator, 3 speed
1922	55646	12" Emerson ventilator, 3 speed
1922	55648	16" Emerson ventilator, 3 speed
1922	Junior	9" Emerson Junior, one speed, no switch, *first junior series
1922	44D	8" Northwind, 2 speed, non-osc
1922	450D	10" Northwind, 3 speed, oscillator
1923	29045	9" Emerson, 3 speed, oscillator, Parker blades
1923	56048	16" Emerson ventilator, 1 speed
1923	56648	16" Emerson ventilator, 1 speed
1923	56666	12" Emerson ventilator, 1 speed
1923	71666	12" Emerson, 3 speed, oscillator, residential, Parker Blades
1923	Junior	9" Ventilator model, Parker blades
1923	450F	10" Northwind, 3 speed, osc

Date	Type	Description and important details
1924	28045	9" Emerson, 3 speed, non-osc desk model, Parker blades
1924	29046	12" Emerson, 3 speed, oscillator, Parker Blades
1924	29048	16" Emerson, 3 speed, oscillator, Parker blades
1924	29266	12" Emerson, 3 speed, oscillator, residential, Parker Blades
1924	44F	8" Northwind, 2 speed, non-osc
1925	28046	12" Emerson, 3 speed, non-osc, Parker blades
1925	Junior	9" Emerson Junior, intorduction of slide switch, shaded pole induction motor
1925	44G	8" Northwind, 2 speed, non-osc
1925	450G	10" Northwind, 3 speed, osc
1925	450H	10" Northwind, 3 speed, osc
1926	44H	8" Northwind, 2 speed, non-osc
1926	444A	8" Northwind, 2 speed, non-osc
1926	450J	10" Northwind, 3 speed, osc
1927	444B	8" Northwind, 2 speed, osc
1927	450K	10" Northwind, 3 speed, osc
1928	55A	8" Northwind, 2 speed, non-osc
1928	450L	10" Northwind, 3 speed, osc
1929	Junior	9" Emerson Junior, S/S, non-osc, Parker blades Style #2581
1929	Junior	9"' Emerson Junior, S/S, non-osc, Parker blades Style #2593
1929	Junior	12" Emerson Junior, S/S, osc, Parker blades Style #2208
1929	28645	12" Emerson, B/S, 3 speed, non-osc, painted Parker blades
1929	29646	12" Emerson, B/S, 3 speed, osc, painted Parker blades, 110v
1929	29646	12" Emerson, B/S, 3 speed, osc, painted Parker blades, 220v
1929	29266	12" Emerson, B/S, 25 cy, 3 speed, osc, painted Parker blades
1929	73648	16" Emerson,B/S, 3 speed, osc, painted Parker blades
1929	73248	16" Emerson,B/S, 25 cy, 3 speed, osc, painted Parker blades
1929	71666	Emerson, B/S, 3 speed, osc, residential fan, painted Parker blades
1929	73668	16" Emerson, B/S, 3 speed, osc, painted Parker blades
1929	71666-FG	12" Emerson,B/S, 3 speed, osc, residential fan in French Grey finish
1929	28045	9" Emerson, B/S,DC, 3 speed, non-osc, painted Parker blades
1929	28046	12" Emerson, B/S, DC, 3 speed, non-osc, painted Parker blades
1929	29045	9" Emerson, B/S, DC, 3 speed, osc, painted Parker blades
1929	75046	12" Emerson, B/S, DC, 3 speed, osc, painted Parker blades
1929	75048	16" Emerson, B/S, DC, 3 speed, osc, painted Parker blades
1929	55	8" Northwind, 2 speed, non-osc
1929	450M	10" Northwind, 3 speed, osc
1934	Seabreeze	8" Seabreeze, S/S, 1 speed, non-osc, ivory color
1934	2240	8" Emerson, S/S, 1 speed, osc, Parker blades
1934	2150	10" Emerson, S/S, 1 speed, non-osc, Parker blades
1934	2250	10" Emerson, S/S, 1 speed, osc, Parker blades
1934	28648	12" Emerson, S/S, 3 speed, non-osc, Parker blades
1934	29646	12" Emerson, S/S, 3 speed, osc, Parker blades
1934	73648	16" Emerson, S/S, 3 speed, osc, Parker blades
1934	71666	12" Emerson, S/S, 3 speed, osc, residential, Parker blades
1934	71666-FG	12" Emerson, 3 speed, osc, residential in French Grey finish, Parker blades
1934	73668	16" Emerson, 3 speed, osc, Parker blades
1934	74045	10" Emerson, DC, 3 speed, non-osc, Parker blades

Date	Type	Description and important details
1934	28046	12" Emerson, DC, 3 speed, non-osc, Parker blades
1934	71045	10" Emerson, DC, 3 speed, osc, Parker blades
1934	75046	12" Emerson, DC, 3 speed, osc, Parker blades
1934	75048	16" Emerson, D.C. 3 speed, osc, Parker blades
1934	??	10" Silver Swan, overlapping blades, osc, 1 speed
1936	7250	10" Silver Swan, overlapping blades, osc, 1 speed
1937	3140-B	8" Seabreeze, non-osc, lily base, Parker blades
1937	3150	10" Seabreeze, non-osc, lily base, Parker blades
1937	3250	10" Seabreeze, osc, lily base, Parker blades
1937	3250-AB	10" Seabreeze, floor fan, osc,
1937	3460	10" Seabreeze, osc, lily base, Parker blades
1937	3350	10" Seabreeze ventalator, window mount
1937	3350-AA	10" Seabreeze ventalator wall box (window)
1937	5460-AA	12" Silver Swan, floor fan, 3 speed, overlapping blades
1937	6250-AA	10" Emerson, 1 speed, overlapping, osc
1937	77646-AK	12" Emerson, 3 speed, overlapping, osc
1937	77648-AK	16" Emerson, 3 speed, overlapping, osc
1937	73646	12" Emerson, 3 speed, osc, Parker blade
1937	73648	16" Emerson, 3 speed, osc, Parker blade
1937	S60-SW	24" Emerson, air circulator, floor fan, non-osc
1937	S60-WB	30" Emerson, air circulator, floor fan, non-osc
1937	D60-DL	24" Emerson, DC air circulator, floor fan, non-osc
1937	D60-TM	30" Emerson, DC air circulator, floor fan, non-osc
1937	2250-B	10" Emerson, 1 speed, osc, chrome finish
1937	73646-AK	12" Emerson, 3 speed, osc, chrome finish
1937	73648-AK	16" Emerson, 3 speed, osc, chrome finish
1937	75046	12" Emerson, DC , 3 speed, osc, chrome finish
1937	75048	16" Emerson, DC, 3 speed, osc, chrome finish
1937	S60WC	24" Emerson, air circulator, 4 blade, chrome finish
1937	S60WA	24" Emerson, air circulator, 3 blade, chrome finish
1937	D60TL	30" Emerson, air circulator, 3 blade, chrome finish
1937	D60TM	24" Emerson, air circulator, 3 blade, chrome finish
1937	5460	12" Silver Swan, 3 speed, osc, overlapping blades
1937	5250-B	10" Silver Swan, 1 speed, overlapping blades, osc
1937	7250	10" Silver Swan-Imperial, 1 speed, all chrome body, overlapping blades
1937	6250	10" Emerson, 1 speed, osc, overlapping blades,
1937	2240-B	8" Emerson, 1 speed, osc, Parker blades
1937	2150	10" Emerson, 1 speed, non-osc, Parker blades
1937	2250-B	10" Emerson, 1 speed, osc, Parker blades
1937	76045	10" Emerson, DC, 1 speed, non-osc, Parker blades
1937	71045	10" Emerson, DC, 1 speed, osc, Parker blades
1937	28646	12" Emerson, 3 speed, non-osc, Parker blades
1937	73646-AK	12" Emerson, 3 speed, osc, Parker blades
1937	73648-AK	16" Emerson, 3 speed, osc, Parker blades
1937	73246	12" Emerson, 25 cy, 3 speed, osc, Parker blades
1937	73248-AA	16" Emerson, 25 cy, 3 speed, osc, Parker blades
1937	28046	12" Emerson, DC, 3 speed, non-osc, Parker blades
1937	75046	12" Emerson, DC, 3 speed, osc, Parker blades
1937	75046	12" Emerson, DC, 230v, 3 speed, osc, Parker blades
1937	75048	16" Emerson, DC, 3 speed, osc, Parker blades
1937	75048	16" Emerson, DC, 230v, 3 speed, osc, Parker blades

Date	Type	Description and important details
1938	3140-B	8" Seabreeze, 1 speed, non-osc, lily base, Parker blades
1938	3150	10" Seabreeze, 1 speed, osc, lily base, Parker blades
1938	3250	10" Seabreeze, 1 speed, osc, lily base, Parker blades
1938	3250-AD	10" Seabreeze, 1 speed, floor fan, osc, lily base, Parker blades
1938	3160	12" Seabreeze, 2 speed, non-osc, lily base, Parker blades
1938	3460	12" Seabreeze, 2 speed, osc, lily base, Parker blades
1938	3460-AB	12" Seabreeze, floor fan, osc
1938	3350	10" Seabreeze ventilator
1938	3350-AA	10" Seabreeze ventilator, box mount
1938	3360	12" Seabreeze ventilator
1938	3360-AA	12" Seabreeze ventilator, box mount
1938	5250-D	10" Silver Swan, 1 speed, osc, overlapping blades
1938	5460-C	12" Silver Swan, 3 speed, osc, overlapping blades
1938	77646-AD	12" Emerson, floor fan, 3 speed, osc, overlapping,
1938	2240-B	8" Emerson, 1 speed, osc, Parker blades
1938	2150	10" Emerson, 1 speed, non-osc, Parker blades
1938	2250-B	10" Emerson, 1 speed, osc, Parker blades
1938	78045	10" Emerson, DC, 1 speed, non-osc, Parker blades
1938	73045	10" Emerson, DC, 1 speed, osc, Parker blades
1938	2250-B	10" Emerson, 1 speed, osc, all chrome finish, Parker blades
1938	72646-AK	12" Emerson, 3 speed, non-osc, Parker blades
1938	73646-AK	12" Emerson, 3 speed, osc, Parker blades
1938	73648-AK	16" Emerson, 3 speeds, osc, Parker blades
1938	28046	12" Emerson, DC, 3 speed, non-osc, Parker blades
1938	75046	12" Emerson, DC, 3 speed, osc, Parker blades
1938	75048	16" Emerson, DC, 3 speed, osc, Parker blades
1938	6250	10" Emerson, 1 speed, osc, overlapping blades
1938	6250-AA	10" Emerson, floor fan, osc, overlapping blades
1938	77646-AK	12" Emerson, 3 speed, osc, overlapping blades
1938	77646-AD	12" Emerson, floor fan, osc, overlapping blades
1938	77648-AK	16" Emerson, 3 speed, osc, overlapping blades
1938	77045	10" Emerson, DC, osc, overlapping blades
1938	77046	12" Emerson, DC. 3 speed, osc, overlapping blades
1938	7250	10" Silver Swan-Imperial, 1 speed, osc, chrome finish
1938	5460-C	12" Silver Swan, 3 speed, osc, overlapping blades
1938	4560-AA	12" Silver Swan, 3 speed, floor fan, osc, overlapping blades

Date	Type	Description and important details
1939	73646-AK-G	12" Emerson, 3 speed, Imp Parker blades, cross guard, osc
1939	77648-AH	16" Emerson, 3 speed, osc, floor fan, cross guard, overlapping
1939	6250-B	10" Emerson, 1 speed, osc, cross guard, overlapping blades
1939	6250-AA	10" Emerson, 1 speed, osc, floor fan, cross guard, overlapping blades
1939	77646-AG	12" Emerson, 3 speed, osc, cross guard, overlapping blades
1939	77646-AD	12" Emerson, 3 speed, osc, cross guard, overlapping blades
1939	77648-AG	16" Emerson, 3 speed, osc, cross guard, overlapping blades
1939	77648-AH	16" Emerson, 3 speed, osc, cross guard, overlapping blades
1939	77646-AK	12" Emerson, 3 speed, cross guard, chrome finish, overlapping blades
1939	77648-AK	16" Emerson, 3 speed, cross guard, chrome finish, overlapping blades
1939	77045	10" Emerson, DC, osc, cross guard, overlapping blades
1939	77046-AA	12" Emerson, DC. osc, cross guard, overlapping blades
1939	5250-D	10" Silver Swan, 1 speed, osc
1939	7250	10" Silver Swan, 1 speed, osc, desk fan
1939	5460-C	12" Silver Swan, 3 speed, osc, desk fan
1939	5250-AA	12" Silver Swan, 3 speed, floor fan, osc
1939	5460-AA	12" Silver Swan, 3 speed, floor fan, osc, overlapping blades
1939	2450-B	10" Emerson, 1 speed, osc, silver painted Parker blades, Rounded "S"
1939	78045	10" Emerson, DC,1 speed, osc, silver painted Parker blades
1939	73045	12" Emerson 3 speed, non-osc, Parker blades
1939	72646-AK	12" Emerson, 3 speed, osc, Parker blades
1939	72648-AK	16" Emerson, 3 speed, osc, Parker blades
1939	73648-AK	16" Emerson, 3 speed, non-osc, Mesh guard, Parker blades
1939	73246	12" Emerson, 25cy, 3 speed, osc, Parker blades
1939	73246-AA	16" Emerson, 25 cy, 3 speed, osc, Parker blades
1939	28046	12" Emerson, DC, 3 speed, non-osc, Parker blades
1939	75048	16" Emerson, DC, 3 speed, osc, Parker blades
1939	75046	12" Emerson, DC 3 speed, osc, Parker blades
1940	5250-D	10" Silver Swan, 1 speed, osc, overlapping blades
1940	7250-B	10" Silver Swan-Imperial, 1 speed, osc, overlapping blades
1940	5460-C	12" Silver Swan, 3 speed, osc, overlapping blades
1940	5250-AA	12" Silver Swan, floor fan, small base, osc, mohogany wrinkle finish
1940	5460-AD	12" Silver Swan, floor fan, small base mahogany wrinkle finish

General Electric

General Electric was founded by Thomas Edison on October 15, 1878, and was called the Edison Electric Light Company. Edison obtained capitol of $300,000 from Grosvenor P. Lowery, a New York City attorney for the Western Union Telegraph Company. Edison was very interested in electricity and felt it was the power source of the future.

In 1889 the Edison Electric Light Company was consolidated with several other Edison companies to form the Edison General Electric Company. Edison was especially interested in direct current applications and did not have much use for alternating current technology. When Westinghouse won the contract to put generators at the Niagara Falls power plant using alternating current technology, the Edison General Electric Company's fortune began to dim.

On April 15, 1892, Edison General Electric and the Thomas-Houston Company merged to form the General Electric Company or GE. In 1894, GE manufactured its first electric desk fan with much success. This fan had a small alternating current motor with no brushes or manual starting mechanism. The fan also incorporated a brass guard to protect the blades from damage. By the late 1890s GE was mass producing high quality desk fans for both alternating and direct current. This trend was to continue for almost a century.

In 1899, GE first used the universally known GE emblem on a fan. This was to be used on all of their products from this time on. Early GE fans had large, flat motors commonly called "pancake" fans. This style of fan was manufactured until about 1907 when a new tank-type motor was introduced.

In 1909 GE introduced the first of several mechanical oscillators. This first large oscillator is referred to as a sidewinder due to the configuration of the levers on the side of the motor. In 1911 and in 1914 GE introduced oscillators that were not met with much success. By the late teens they had settled on an oscillator that was used for several decades.

By the twenties GE was manufacturing large quantities of electric desk fans. Fan catalogs of the period show a large variety of fans available for consumer consumption. These fans were of high quality, but lagged behind technologically what Emerson was manufacturing at the same time.

The Depression changed the way fans were manufactured and designed drastically. General Electric seemed to understand this more than any other fan manufacturer and they came out with newly designed fans first. In 1931 they introduced the 10" Quiet Fan with large overlapping highly polished aluminum blades. The following year they put the new quiet overlapping blades on several of the trade fans, thus leading all other manufacturers in this trend.

In 1938 GE introduced the Vortalex trailing tip blade, another design innovation. This new blade design allowed the trailing vortex produced by the moving blade to cut the air more rapidly and quietly and allowing for a better air flow. This new design gave GE fans a distinct look and became a hallmark of their post-World War II fans.

Following World War II General Electric manufactured many Vortalex fans. In 1947 they came out with a newly designed fan with a cast-plastic body and a new oscillator which was located at the front of the fan.

In 1984 General Electric sold their small motor division to Black & Decker, thus ending a century of fan manufacturing.

General Electric first began production of electric desk fans in 1894. These fans were very modern in design and had several standard features such as a brass guard and no internal starting switch. The early General Electric fans had six blades, but by 1899 they had four blades.

Left: ***GE***
Type: unknown
Blades: 12"
Date: 1899

Right: ***GE***
Type: 265731
Blades: 12"
Date: 1906

This fan was one of the first to use the GE logo on its badge and is a prototype of early General Electric fans. The base is made of cast iron and forged to the motor. The motor is made of cast iron with large vent holes located on the front and rear of the motor. A two-speed rotary switch is located at the center of the back of the motor. Two terminals are located below the switch and are attached to a power source. A brass strip goes completely around the center of the motor which contains information about the fan. A brass ten S wire guard is attached to the motor by four brass struts. The blades are made of heavy sheet brass in the windmill style and are attached to a cast-brass hub. The motor has a single bearing and is very quiet to operate.

Grade I – $750.00
Grade II – $563.00
Grade III – $375.00
Grade IV – $263.00
Grade V – $150.00
Grade VI – $38.00

A classic example of a very late pancake fan. The castings are very heavy and smooth with no ribbing or decoration. The base contains a five-speed rotary switch. The motor is attached to the base by a large cast-iron trunnion. The pancake motor has large air vents and exposed terminals on the back of the fan motor. The brass eight S wire guard is attached to the motor by three brass struts. The blades are made of heavy sheet brass attached to a brass spider that is open in the front. A large brass oil cup is located just under the front bearing.

Grade I – $350.00
Grade II – $263.00
Grade III – $175.00
Grade IV – $123.00
Grade V – $70.00
Grade VI – $18.00

In 1903 GE produced a series of fans for about seven months. It seems that General Electric was looking for the right mix of design and cost. This group of fans has a full ribbed base and may have ten-wire or eight-wire guard. They are unusual because of the complete ribbing on the base.

Left: ***GE***
Type: AK-Ford D, No. 181327
Blades: 12"
Date: 1903

This fan has a ten S wire guard attached to a pancake motor by four brass struts. The motor is made of cast iron held to the base by a trunnion. There are exposed terminals at the back of the motor. The base has a five-speed rotary switch and is fully ribbed.

Grade I – $500.00
Grade II – $375.00
Grade III – $250.00
Grade IV – $175.00
Grade V – $100.00
Grade VI – $35.00

Right: ***GE***
Type: unknown
Blades: 16"
Date 1903

A 16" pancake fan with an eight S wire guard attached to the motor by four brass struts. The blades are made of heavy sheet brass attached to a brass spider that is open in the front. The motor has both front and rear bearings, the rear bearing being covered by a small brass cap stamped "GE." The motor sits on a cast-iron trunnion attached to a full ribbed base. The base contains a five-speed rotary switch.

Grade I – $600.00
Grade II – $450.00
Grade III – $300.00
Grade IV – $210.00
Grade V – $120.00
Grade VI – $30.00

General Electric also produced many direct current fans. Thomas Edison was very keen on direct current and would have preferred it over alternating current. General Electric made their first direct current fans in a pancake style, but soon moved to a more modern style as pictured at right.

Right: ***GE***
Type: D
Blades: 12"
Date: c.1903

A very sleek design having completely hidden brushes. The brushes are hidden by a plate that covers the rear of the motor that can be easily removed for cleaning and maintenance. The motor has two bearings with large brass grease cups. The rear bearing is covered with a stamped brass GE cap. The motor is attached to the base by a cast-iron trunnion. The base is fully ribbed and very similar to Emerson bases of the period. The guard is made of formed brass wire and is attached to the motor by four brass struts. The blades are made of heavy sheet brass attached to a brass spider open in the front.

Grade I – $700.00
Grade II – $525.00
Grade III – $350.00
Grade IV – $245.00
Grade V – $140.00
Grade VI – $35.00

Right: A cast-iron wall mount bracket for a direct current fan. Early fans were not flexible in the sense of being able to be placed on a table or mounted on a wall without being taken apart and modified. The consumer had to order a special bracket to fit the fan to place it on the wall. These are commonly called bracket fans. Most of these fans were for commercial or industrial use and marketed as such.

Price: $175.00

Early desk fans were not flexible enough to be mounted on the wall so manufacturers offered special brackets that could make this possible. Bracket fans were modified desk fans that could be placed on the wall. They used special brackets to hold the motor in a fixed location on the wall. Bracket fans were popular on railway cars and street cars at the turn of the century. Manufacturers soon began to modify their desk fans for this application.

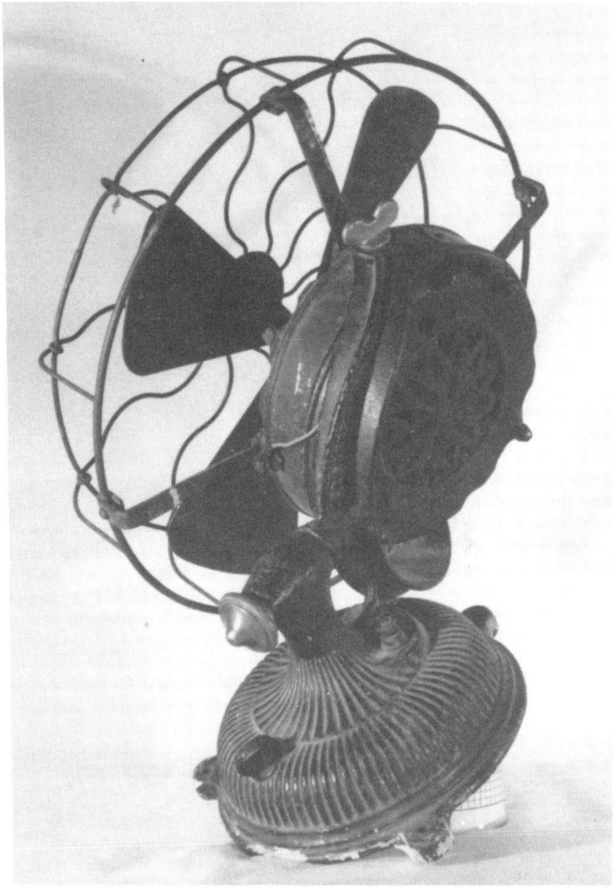

GE
Type: 17239
Blades: 12"
Date: 1903

This example has a full ribbed base with a five-speed rotary switch located at the bottom of the base (when attached to the wall). A pancake motor is attached to the bracket by a trunnion. There are two exposed terminals at the back of the motor which were to be attached to the bracket and then to a power source. The guard is made of formed brass with ten S wires attached to the motor by four brass struts. The blades are made of heavy sheet brass in the windmill style. Note the ornate casting with large vent holes on the back of the motor, this is characteristic of many pancake motors and served to help cool the motor.

Grade I –	$800.00	Grade IV –	$280.00
Grade II –	$600.00	Grade V –	$160.00
Grade III –	$400.00	Grade VI –	$60.00

Left: ***GE***
Type: unknown
Blades: 12"
Date: 1905

Right: ***GE***
Type: unknown
Blades: 12"
Date: 1901

In 1905 GE introduced this newly designed fan for consumer use. There were several modifications to this fan from the older pancake designs, one being that the guard had eight wires instead of ten. The construction of the guard was improved by having the front S wires attached to the back support wire by being wrapped around instead of being placed into drilled holes in the ring. The guard is attached to the motor by three brass struts instead of four. The motor is attached to the base by a large trunnion. The base is half ribbed and contains a five-speed rotary switch. This picture shows the small "off" brass plate located at the base of the switch. These were produced in great quantities.

Grade I – $350.00
Grade II – $263.00
Grade III – $175.00
Grade IV – $123.00
Grade V – $70.00
Grade VI – $18.00

In 1901 GE introduced a newly designed fan that had an adjustable motor that could be moved up or down by use of a trunnion. These fans were also marketed in a non-trunnion model. The base is cast iron with partial ribbing. The pancake motor is made of cast iron and has a two-speed rotary switch on the back. Two exposed terminals are located below the switch. There are no electrical parts in the base. The guard is a brass ten S wire attached to the motor by four brass struts. The blades are made of heavy sheet brass and windmill in style

Grade I – $700.00
Grade II – $525.00
Grade III – $350.00
Grade IV – $245.00
Grade V – $140.00
Grade VI – $ 35.00

In 1908 General Electric made many major changes on their desk fans. The pancake motor was discontinued for a sleeker, more compact motor. The five-speed rotary switch was replaced by a three-speed slide switch located at the front of the base. The windmill type blade was replaced by a rounded, quieter blade. All electric terminals were placed inside the motor and not exposed as on the pancake motor. This was the first truly modern fan made by General Electric.

Left: ***GE***
Type: 361302
Blades:16"
Date: 1908

A prototype of General Electric fans to come. The castings are very heavy making it difficult to transport. The slide switch is rather hard to move back and forth, a long-time problem with GE switches. The motor is held to the base by a trunnion. The newly designed motor is made of cast iron with front and back bearings. Some of the fans produced at this time had internal starting switches. The fan pictured above does not. The guard is made of brass and is attached to the motor by three brass struts. In this design the guard tends to move side by side if the machine nuts are not tightly affixed to the motor. This defect was corrected the following year.

Grade I – $300.00
Grade II – $250.00
Grade III – $150.00
Grade IV – $106.00
Grade V – $60.00
Grade VI – $15.00

Right: ***GE***
Type: 905585
Blades: 12"
Date: 1909

The major improvement of this fan over the 1908 model is that the guard is attached to the motor in a much more secure fashion. The engineers at GE improved the guard struts by placing a bend in the end that attaches to the motor. This change was used on GE fans for decades which enabled the guard to be much sturdier. The castings are very heavy, making the fan awkward to move. The base has numbers painted above the switch to indicate speed or off.

Grade I – $250.00
Grade II – $188.00
Grade III – $125.00
Grade IV – $53.00
Grade V – $30.00
Grade VI – $12.00

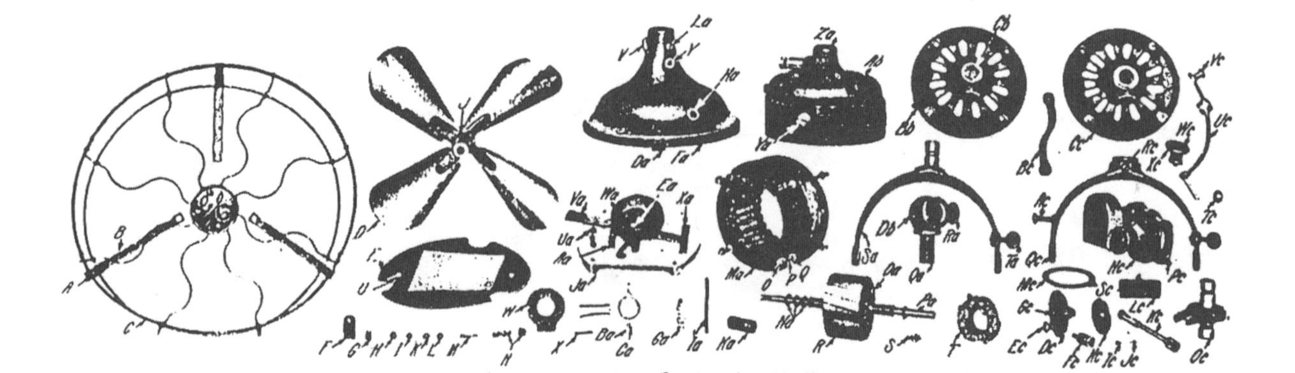

A Guard Clamping Screw
‡ B Guard Support with Clamp and Screw
‡ C Fan Guard, complete
‡ D Fan, complete with Set Screw
E Fiber Base Plate
F Armature Shaft Bearing
G Oil Wick Collar
H Nut for Switch Contact Plug
I Washer for Switch Contact Plug
J Set Screw for Fan
K Washer for Regulator Coil Screw
L Nut for Regulator Coil Screw
M Spring Collar for Switch Blade Pivot
N Switch Connection, complete
O Cap Nut for Field Stud
P Field Stud
Q Clamping Nut for Field Stud
R Armature, complete (State whether Standard or Oscillating)
S Centrifugal Switch Insulated Screw
T Centrifugal Switch, complete

U Screw for Base Plate
V Thumbscrew for Base and Yoke
W Insulating Plate for Collector Ring Leads
X Screw for Switch Base
Y Soft Rubber Bushing for Field Lead
Aa Switch Ratchet
Ba Collector Rings with Insulating Plate
Ca Screw for Collector Rings
Da Soft Rubber Foot for Base
Ea Screw for Regulator Coil
‡ Fa Motor Base with Screws and Bushings
Ga Oil Wick Spring
Ha Soft Rubber Bushing for Attaching Cord
Ia Oil Wick
Ja Regulator Switch, complete less Regulator Coil
Ka Oil Cup
La Set Screw for Base and Yoke
Ma Field, complete
Na Shaft Washers
Oa Spacing Block for Centrifugal Switch

Pa Armature Shaft (State whether Standard or Oscillating)
Qa Angular Adapter
Ra Adapter Thumbscrew
Sa Yoke for Motors except Oscillating
Ta Yoke Thumbscrew
Ua Switch Contact Plug
Va Switch Blade
Wa Regulator Coil
Xa Switch Contact Plate
Ya Trunnion Screw
Za Motor Frame
Ab Guard Support Screw
Bb Cap for Motor except Oscillating
Cb Cap Monogram for Motors except Oscillating
Db Adapter Set Screw
*Ac Stud for Rocker Arm
*Bc Connecting Rod
*Cc Cap for Oscillating Motor

*Dc Spur Gear
*Ec Screw for Spur Gear & Connecting Rod
*Fc Worm
*Gc Screw with Nut for Spur Gear
*Hc Worm Gear
*Ic Locking Pin for Operating Knob.
*Jc Compression Spring for Operating Knob
*Kc Gear Spindle with Pinion
*Lc Gear Case Lid
*Mc Shield for Spur Gear
*Nc Gear Case
*Oc Swivel
*Pc Screw Fastening Gear Case in position
*Qc Yoke for Oscillating Motor
*Rc Set Screw for Yoke and Swivel
*Sc Screw for Gear Case Lid
*Tc Ball for Rocker Arm
*Uc Rocker Arm
*Vc Pivot Screw for Rocker Arm
*Wc Operating Knob, complete
*Xc Set Screw for Operating Knob

‡ State whether for 12 or 16 inch Fan
* For Oscillating Motor only

In Ordering, give name of part, reference letter, photograph number (303869) and Serial number of Motor

An illustration from a 1909 General Electric fan catalog with replacement parts for a 12" fan. Not all of the parts are interchangeable and had to be reviewed carefully when ordering. There are two trunnions listed, one for an oscillator, the other a non-oscillator. This fan would have had the sidewinder oscillator. It is interesting to note that the armature has a place for a centrifugal starting switch. All fans I've seen of this period that were osicllators did not have a centrifugal starting switch.

In 1909 General Electric introduced their first mechanical oscillator. This new device was located at the rear of the fan and had a lever that extended out from the side of the gearbox attached to the top of the trunnion. This unusual configuration is referred to as a sidewinder oscillator. The castings are massive making the fan look rather awkward. A brass button on the side of the oscillator would engage the gearbox when oscillation was desired. This design was met with modest success and was soon replaced.

Left: ***GE***
Type: 398950 Sidewinder Oscillator
Blades: 12"
Date: 1909

This is an unusual example of a sidewinder oscillator because it is a residential six-blade fan. Note how heavy the castings are. The trunnion is used to move the fan up or down.

Grade I –	$2,400.00
Grade II –	$1,800.00
Grade III –	$1,200.00
Grade IV –	$840.00
Grade V –	$480.00
Grade VI –	$120.00

Right: ***GE***
Type: 427220 Sidewinder Oscillator
Bladcs: 16"
Date: 1909

The taller base on this example makes the fan look more modern than its 12" counterpart. The blades are made of heavy sheet brass and the guard is brass. This fan is too large and heavy not have a handle for transport. You can see the oscillator button located at the rear of the gear box.

Grade I –	$3,600.00
Grade II –	$2,700.00
Grade III –	$1,800.00
Grade IV –	$1,260.00
Grade V –	$720.00
Grade VI –	$180.00

Left: **GE**
Type: 342864
Blades: 12"
Date: 1908

Left: **GE**
Type: 624422 Kidney Oscillator
Blades: 12"
Date: 1911

An example of the first non-pancake fan with a newly designed motor. This example is unusual because it has an internal starting switch, not a common feature found on General Electric fans. The blades are made of thick sheet brass and are deeply pitched, another unusual characteristic. The base is large and squatty. Note the three brass guard struts are flat and not bent at the point where they are affixed to the motor housing. This is an important characteristic of the 1908 model.

Grade I – $350.00
Grade II – $263.00
Grade III – $175.00
Grade IV – $123.00
Grade V – $70.00
Grade VI – $18.00

This example has the characteristic oscillator ring around the motor so commonly found on later General Electric products. The castings are very heavy and finished in a gloss black paint. The blades are made of heavy sheet brass and highly polished. The guard is made of formed brass wire in the S style. Note how the bottom guard struts are formed to fit around the oscillator ring, a configuration that became standard on General Electric oscillators.

Grade I – $500.00
Grade II – $375.00
Grade III – $250.00
Grade IV – $175.00
Grade V – $100.00
Grade VI – $35.00

Left: ***GE***
Type: 861916 Kidney Oscillator
Blades: 12"
Date: c.1913

Right: ***GE***
Type: 75423 Star Oscillator
Blades: 12"
Date: c.1918

This is a late example of a kidney oscillator with an improved modified oscillator ring with a handle. This handle became a standard feature on General Electric oscillators from this time forward. This is an unusual example because it is a six-blade residential fan. The blades are made of heavy sheet brass finished in polished brass. The blades push out from the motor in a strange configuration making the fan look awkward. The guard is made of formed brass wire in the S style and protrudes forward from the motor housing. The castings are very heavy and finished in a gloss black.

Grade I – $500.00
Grade II – $375.00
Grade III – $250.00
Grade IV – $175.00
Grade V – $100.00
Grade VI – $35.00

Around World War I General Electric began using a new improved gear box which had an unusual brass oscillator set screw that was formed in the configuration of a star. These fans were the forerunner of the commonly produced oscillators of the twenties and early thirties. This example has a brass star set screw at the base of the oscillator wheel. The castings are very heavy and finished in a gloss black. The blades and guard are made of brass. This fan has three speeds.

Grade I – $250.00
Grade II – $188.00
Grade III – $125.00
Grade IV – $53.00
Grade V – $30.00
Grade VI – $12.00

Residential six-blade fans became very popular around 1910, and were made by several manufactures. The reasoning behind the residential fan was that the blade rotated at a slower rate making them quieter for home use. Residential fans are very desirable and sought after by collectors because of this different configuration.

Left: ***GE***
Type: 89652 Star Oscillator
Blades: 12"
Date: c.1917

An example of a continuous star oscillator with brass blades and guard. A continuous oscillator will sit in a stationary position, but move back and forth when operating. The castings are very heavy and have a handle formed into the top of the oscillator ring. The body of the fan is finished in gloss black. This fan has three speeds.

Grade I – $450.00
Grade II – $338.00
Grade III – $225.00
Grade IV – $158.00
Grade V – $90.00
Grade VI – $22.00

Center: ***GE***
Type: 568752
Blades: 12"
Date: c.1910

An early example of a residential fan with a tall base mounted on a trunnion. It seems that General Electric experimented with several base styles around 1910. This may have been because in the preceding year they had introduced their first mechanical oscillator. The older squatty base seemed to be inappropriate for oscillating fans that had a large gear case protruding from the rear of the fan. This taller base gave the fan a more dignified and streamlined look.

Grade I – $500.00
Grade II – $375.00
Grade III – $250.00
Grade IV – $175.00
Grade V – $100.00
Grade VI – $35.00

Right: ***GE***
Type: 78777 Star Oscillator
Blades: 12"
Date: c.1918

This example is a star oscillator with heavy castings. The blades are made of heavy sheet brass that have a shallow pitch. The guard is made of formed brass wire in the S style. The fan is finished in gloss black and has three speeds.

Grade I – $450.00
Grade II – $338.00
Grade III – $225.00
Grade IV – $158.00
Grade V – $90.00
Grade VI – $22.00

Around 1910 small fans for home use became available on the market. These smaller fans were marketed for use in the home as well as the office. These fans were touted as giving cooling relief to sweltering summer heat while using an inexpensive amount of electricity. During the twenties these small fans made up a large share of the market.

Left: **GE**
Type: 174948
Blades: 9"
Date: c.1917

This small fan has a cast-iron motor housing and base. It has the newer gear box attached to the rear of the fan. The motor is surrounded by an oscillator ring. The blades are made of polished brass. The guard is steel, but is constructed in the older wrap around style. The fan has three speeds and is finished in gloss black.

Grade I – $100.00
Grade II – $75.00
Grade III – $50.00
Grade IV – $35.00
Grade V – $20.00
Grade VI – $5.00

Right: **GE**
Type: 71951
Blades: 9"
Date: c.1916

All brass fans were common before World War I. This example is unusual because it has an oscillator. The fan is made entirely of brass finished in gun blue. The guard and blades are polished brass. The switch and field coils are tightly packed into the base and motor housing of this very compact fan. Note that the motor badge is located on the side of the motor, that was a common characteristic of star oscillators of this period.

Grade I – $250.00
Grade II – $188.00
Grade III – $125.00
Grade IV – $53.00
Grade V – $30.00
Grade VI – $12.00

An illustration of a standard General Elecric fan from 1920. These fans were manufactured in large numbers but were not of the same quality as Emersons of the same period.

Left: ***GE***
Type: unknown
Blades: 8"
Date: c.1927

A small inexpensive home use fan. The base and motor housing are made of stamped steel. The fan is finished in gloss hunter green.

Grade I – $70.00
Grade II – $53.00
Grade III – $35.00
Grade IV – $25.00
Grade V – $14.00
Grade VI – $3.00

Center: ***GE***
Type: 940566 Collar Oscillator
Date: 1914

In 1914 General Electric introduced their third mechanical oscillator. This new design had completely enclosed gears making the fan look very sleek and modern. To engage the oscillator the operator would rotate a brass ring located at the base of the motor. Unfortunately, some of the gears were not sturdy enough to handle the pressures of oscillation and would often break. Though very modern in design, this oscillator was soon abandoned for a much simpler, easy to produce model. This is unfortunate because this oscillator was far ahead of its time in style and operation. The blades and guard are made of polished brass. The motor has been redesigned and is very modern looking.

Grade I – $350.00
Grade II – $263.00
Grade III – $175.00
Grade IV – $123.00
Grade V – $70.00
Grade VI – $18.00

Right: ***GE***
Type: Ser. G
Blades: 6"
Date: c.1924

This is one of the smallest fans made by General Electric. It has stamped brass blades with a steel guard. The motor housing and base are made of stamped steel finished in gloss hunter green. The motor is AC/DC and is unusual for General Electric fans of the period. This fan also comes in an ivory color.

Grade I – $70.00
Grade II – $53.00
Grade III – $35.00
Grade IV – $25.00
Grade V – $14.00
Grade VI – $3.00

Left: ***GE***
Type: 92880
Blades: 9"
Date: c.1919

This is another all brass fan in a non-oscillator model. The original finish was gun blue, but this fan has been painted. The guard and blades are polished brass. This fan has three speeds.

Grade I – $190.00
Grade II – $143.00
Grade III – $95.00
Grade IV – $67.00
Grade V – $38.00
Grade VI – $10.00

Right: ***GE***
Type: AR
Blades: 9"
Date: c.1920

This is an early small home use fan with a cast-iron base and motor housing. It has two speeds and is a non oscillator. The guard and blades are made of steel. The fan is finished in dark matte green and is an early example of a General Electric green colored fan.

Grade I – $90.00
Grade II – $68.00
Grade III – $45.00
Grade IV – $32.00
Grade V – $18.00
Grade VI – $5.00

As demand for small home use fans grew in the early twenties, General Electric came out with a new line of inexpensive fans call the Whiz. These fans were very attractive and efficient. Unfortunately many of the motor housings were made of die-cast metal that did not hold up to the test of time. The oscillator castings were especially prone to breaking. Some examples had brass blades, other had steel.

Left: ***GE***
Type: 257599 Whiz
Blades: 9"
Date: c.1924

This fan has a newly designed stamped steel base that is sleek and attractive. The traditional slide switch has been replaced by a one-speed toggle switch. The blades are made of polished brass and go nicely with the gloss hunter green paint. This example has an oscillator.

Grade I – $70.00
Grade II – $53.00
Grade III – $35.00
Grade IV – $25.00
Grade V – $14.00
Grade VI – $3.00

Right: ***GE***
Type: 236327 Whiz
Blades: 9"
Date: c.1924

A non-oscillating Whiz finished in gloss hunter green. The blades are polished brass and there is a one-speed toggle switch in the base. These fans were very popular when they were first introduced.

Grade I – $70.00
Grade II – $53.00
Grade III – $35.00
Grade IV – $25.00
Grade V – $14.00
Grade VI – $3.00

Coin operated devices become popular in the late ninetieth century and this application was soon adapted to fans. Both General Electric and Westinghouse made coin operated fans for consumer use. General Electric was by far the most common manufacturer and these are two examples. The fan would give the consumer an hour of use for each nickel deposited. The more nickels deposited, the longer the fan would operate. This was ideal for hotels in warmer climates. The castings are very ugly and massive. A small handle has been attached to the motor for greater mobility.

Left: ***GE***
Type: 272036-1 Coin-Op
Blades: 12"
Date: c.1915

This example is for alternating current. The base is made of cast iron and is very heavy. There is a slot located at the top of the base to deposit a coin. The large fluted knob on the back of the base has instructions how to operate the fan. The hole at the base of the fan was the bank which held the nickels, the door on this example is missing. The motor is the 1914 collar oscillator design, and this fan does not oscillate. The blades are made of polished brass, but later examples are stamped steel. A very unusual fan.

Grade I – $300.00
Grade II – $250.00
Grade III – $150.00
Grade IV – $106.00
Grade V – $60.00
Grade VI – $15.00

Right: ***GE***
Type: 272036-1 Coin-Op
Blades: 12"
Date: c.1915

This is a direct current example. Note the brush caps at the back of the motor. Being a direct current example is unusual.

Grade I – $400.00
Grade II – $300.00
Grade III – $200.00
Grade IV – $140.00
Grade V – $80.00
Grade VI – $20.00

Left: ***GE***
Type: 34017
Blades: 12"
Date: c.1922

Right: ***GE***
Type: 34021
Blades: 16"
Date: c.1922

This is a very standard early twenties non-oscillating fan with cast-iron base and motor housing. The blades are steel finished in a silver matte color. The guard is made of steel with a badge painted silver to match the blades. The base and guard are finished in dark matte green color. This fan has three speeds.

Grade I – $150.00
Grade II – $112.00
Grade III – $75.00
Grade IV – $53.00
Grade V – $30.00
Grade VI – $8.00

This large fan would have most likely had an industrial application. The blades are made of steel painted silver. The guard is made of formed steel wire in a S style. The silver paint on the badge has mostly worn off but originally matched the blades. The base and motor housing are made of cast iron finished in a matte green color.

Grade I – $170.00
Grade II – $128.00
Grade III – $85.00
Grade IV – $60.00
Grade V – $34.00
Grade VI – $9.00

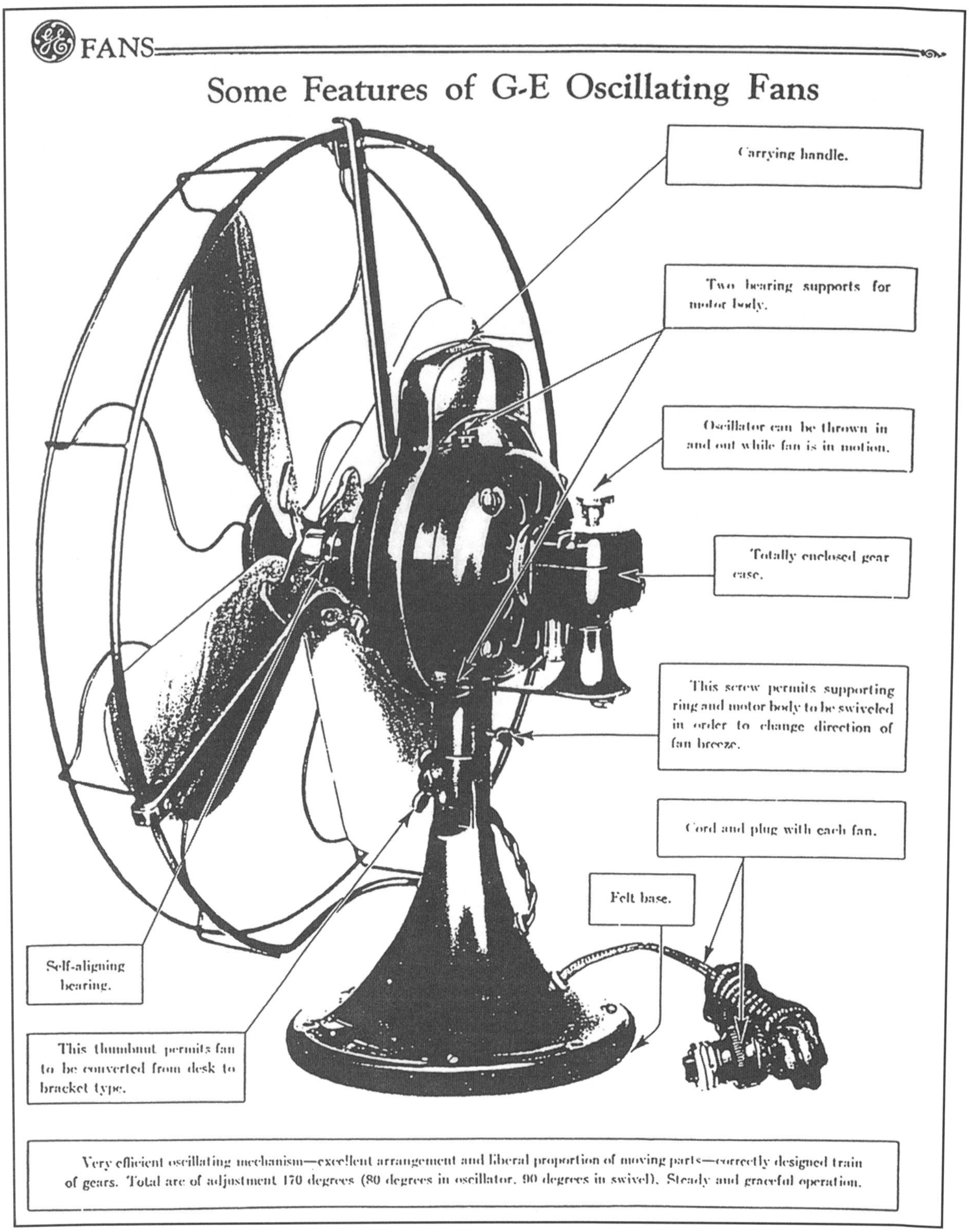

An illustration of a standard General Electric fan from the middle to late twenties. This fan has standard General Electric features such as a carrying handle, oscillator ring around the motor, three speeds, and thin shallow pitched blades. These fans were finished in hunter green and manufactured in large numbers.

By the middle 1920s General Electric was producing large numbers of electric fans for commercial and home use. Standard features on these fans included a hunter green finish, steel blades, and an S type guard. The motor housing was cast iron on the larger models, and stamped steel on the smaller models. Oscillating fans had a cast-iron ring around the motor with a cast handle at the top of the 12" and 16" models. General Electric made many variations of these fans with some models having two or three speeds, slide or toggle switches, and oscillating or non-oscillating. General electric began marketing fans in hunter green as a marketing device because most black fans looked the same. This made GE fans more noticeable on the store shelf. Hunter green became a hallmark of GE fans from the late teens until the early thirties.

Left: ***GE***
Type: 34X899
Blades: 16"
Date: c.1928

A very typical oscillating industrial grade fan with large, shallow blades, a cast-iron oscillating ring around the motor attached to a cone base and a three-speed slide switch. Millions of these fans were manufactured.

Grade I – $90.00
Grade II – $68.00
Grade III – $45.00
Grade IV – $32.00
Grade V – $18.00
Grade VI – $5.00

Center: ***GE***
Type: unknown
Blades: 9"
Date: c.1928

Small home use fans were very popular in the twenties. This example is an oscillator with an open S wire guard. This fan is not the correct color and should be hunter green with a brass badge. The base has a small toggle switch for operation.

Grade I – $30.00
Grade II – $23.00
Grade III – $15.00
Grade IV – $10.00
Grade V – $6.00
Grade VI – $2.00

Right: ***GE***
Type: 19X263
Blades: 10"
Date: c.1928

This example has a two speed slide switch in the base with a continuous oscillator. The two-speeds made this a deluxe model while the more modest fans only had one speed.

Grade I – $50.00
Grade II – $38.00
Grade III – $25.00
Grade IV – $18.00
Grade V – $10.00
Grade VI – $3.00

DETAILS OF THE G-E OSCILLATING MECHANISM

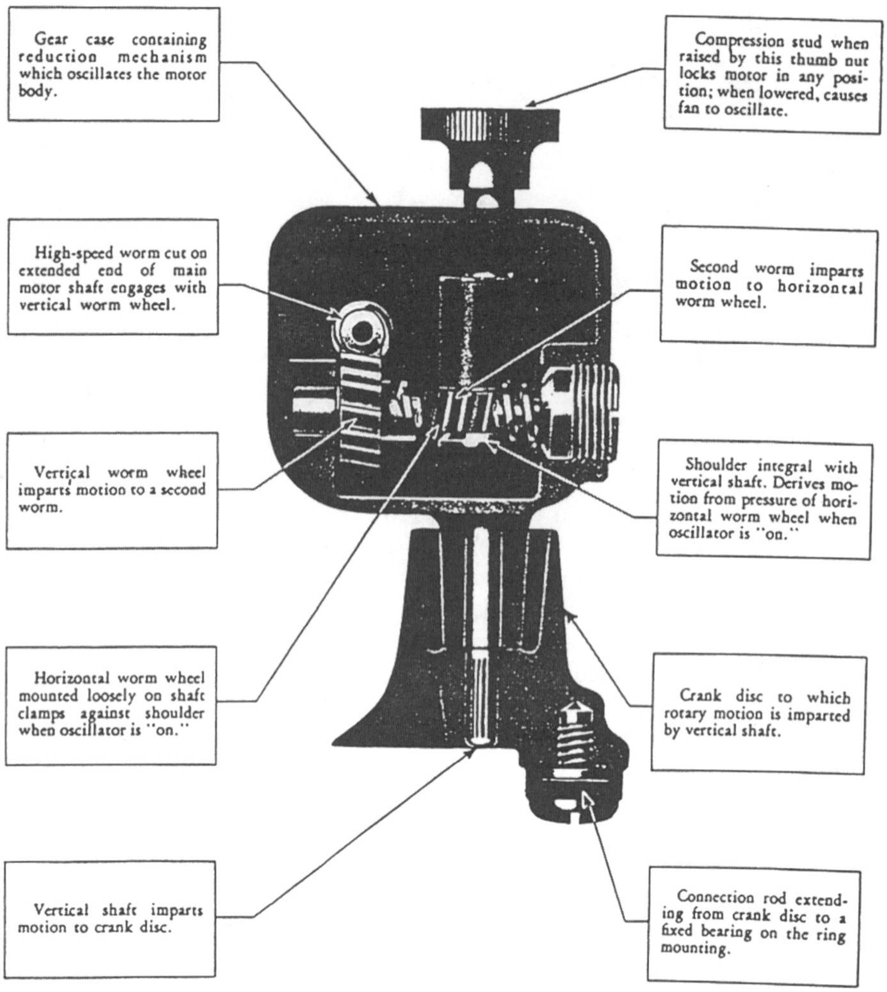

G-E Oscillating Fan Facts

General Electric Oscillating Fans are equipped with an extremely efficient yet simple oscillating mechanism which shifts the fan through an arc of 80 degrees. The oscillator is thrown "in" and "out" by means of the upper adjusting nut. An additional adjustment nut allows the fan body to be swiveled 90 degrees, thus allowing the fan to be adjusted to any point within an arc of 170 degrees.

A two-bearing support for motor body, straight connecting rod and a correctly designed train of gears make possible silent and certain operation throughout the entire range. The entire oscillating mechanism is packed in a high-grade grease which requires renewal only once a season.

An illustration of a General Electric oscillating mechanism used on all fans made after about 1917. This oscillator met with modest success and was inferior to those oscillators made by Emerson and Westinghouse. It was fairly easy to use and would oscillate the fan within an arc of 170 degrees.

All Chrome fans began being produced in the middle twenties and reached their peak of popularity in the thirties. These fans were marketed for commercial use in stores, doctor's offices, drug stores, hospitals, and other places where a very clean appearance was desired. These fans were more expensive than painted varieties and are much rarer. This example is an early model having a brass guard and blades plated in polished chrome.

GE
Type: 75425CP
Blades: 16"
Date: c.1922

An example of an industrial grade fan with a highly polished chrome finish.

Grade I –	$350.00	Grade IV –	$123.00
Grade II –	$263.00	Grade V –	$70.00
Grade III –	$175.00	Grade VI –	$18.00

In 1929 General Electric marketed a new DeLux fan for residential use. This fan was an attempt to make a fan look more like a piece of furniture. The new DeLux fan had a stamped steel body attached to a large, stylized trunnion on a round base. The finish is dark brown mahogany with gold trim. The motor is a standard 10" fan motor that had a multi-bladed blade. These fan were marketed for more affluent consumers and were made for only several years.

Left: **GE**
Type: 19X263
Blades: 10"
Date: c.1928

A very typical home use oscillator with one-speed toggle switch. Compare this fan to the newly designed DeLux fan on the right.

Right: **GE**
Type: 22X424
Blades: 10" DeLux Fan
Date: 1929

A newly designed fan for home residential use. A radical break from the traditional fans of the period.

Grade I –	$700.00	Grade IV –	$245.00
Grade II –	$525.00	Grade V –	$140.00
Grade III –	$350.00	Grade VI –	$35.00

After the stock market crash of 1929 fan sales began to fall sharply. By 1930 profits were very bleak. General Electric began to make drastic design changes in their fans to boost sales. This fan came out in 1931 and stunned the fan manufacturing community because of the newly designed quiet blades. This new design sent shock waves throughout the fan market. Nothing like this had ever been seen before. The results of this new design improved fan sales and changed the look of General Electric fans forever. The following year General Electric began to put the newly designed quiet blades on many of their mass-produced fans.

Right and Left: ***GE***
Type: 33X164 Quiet Fan
Blades: 10″
Date: 1932

Everything about this fan is different and newly designed. The base is made of cast metal in a streamlined designed. The finish is black with silver trim and some later examples were light brown. There is a one-speed toggle switch at the front of the base with "General Electric" cast into it. The motor is almost completely covered by a highly polished aluminum three wing blade. The guard is made of steel wires radiating out from a large center ring. A large black and silver GE badge is located on the back of the motor. Some of these examples have a Graybar tag, but they were all manufactured by General Electric.

Grade I –	$130.00	Grade IV –	$45.00
Grade II –	$98.00	Grade V –	$26.00
Grade III –	$65.00	Grade VI –	$7.00

General Electric did not abandon their regular line of fans after the crash of 1929. They did however make several minor design changes that were mostly cosmetic. The guard was modified using a heavier grade steel wire and a third wire was added for strength and appearance. The rest of the fan was exactly like the fans made in the late twenties.

Left: **GE**
Type: 49X612
Blades: 10"
Date: 1936

This is a deluxe model with a three-speed slide switch in the base. The fan is finished in hunter green with a brass and green GE badge.

Grade I – $60.00
Grade II – $45.00
Grade III – $30.00
Grade IV – $21.00
Grade V – $12.00
Grade VI – $3.00

Right: **GE**
Type: 42X548
Blades: 10"
Date: 1936

An inexpensive home use fan with a one-speed toggle switch. The fan is finished in hunter green with a brass and green GE badge. This fan also came in a non-oscillating model.

Grade I – $40.00
Grade II – $30.00
Grade III – $20.00
Grade IV – $14.00
Grade V – $8.00
Grade VI – $2.00

Following the success of the 1931 Quiet Fan, General Electric began a new line of fans with newly designed quiet aluminum blades. Production of these fans began in 1932, but the patent for this blade was not completed until 1934. The first group of these fans with this new blade were designed just like the earlier fans of the late twenties. These examples are seen on the right and left of the picture. Later, in the middle thirties, GE redesigned the guard as seen in the center of the picture. These quiet blade fans were very successful and brought GE much needed revenue from fan sales.

Left: ***GE***
Type: 55X165B
Blades: 9"
Date: 1932

This fan incorporates only one new feature: newly designed overlapping blades. The traditional S wire guard has been pushed out to make room for the larger blades. The motor, oscillator, base, and switch are all carried over from the 1920s design.

Grade I – $90.00
Grade II – $68.00
Grade III – $45.00
Grade IV – $32.00
Grade V – $18.00
Grade VI – $5.00

Center: ***GE***
Type: 49X723
Blades: 10"
Date: 1936

This fan has a newly designed guard that is much more streamlined in style than the traditional S wire guard. The ring around the motor has been discontinued and the toggle switch in the base has been redesigned.

Grade I – $80.00
Grade II – $60.00
Grade III – $40.00
Grade IV – $28.00
Grade V – $16.00
Grade VI – $4.00

Right: ***GE***
Type: 55X165
Blades: 9"
Date: 1932

During the thirties fans were produced in other colors besides black. This is an example of that trend. The ivory finish makes this very collectible. This fan has all the standard features of earlier fans manufactured by General Electric with exception of large quiet blades.

Grade I – $110.00
Grade II – $83.00
Grade III – $55.00
Grade IV – $39.00
Grade V – $23.00
Grade VI – $6.00

GENERAL ELECTRIC *Quiet* FAN

LARGE OSCILLATING

The large Oscillating Quiet Fan illustrated here is General Electric's most powerful desk-type Quiet Fan. It delivers a tremendous volume of air and distributes it effectively by means of an efficient oscillating mechanism. Especially recommended for offices, stores, restaurants, hotels, theatres and auditoriums.

Outstanding features include the overlapping "Quiet Blades" which move more air than conventional blades of larger diameter, unusually quiet operation, three-speed switch, and attractive appearance.

FEATURES

Motor—Same as motor in 16-in. General Electric conventional model. Self-aligning ball-seat bearings and a special lubricating system prevent wear on bearings and eliminate scoring of shaft by wick. Grease cups contain non-hardening grease sufficient for 1500 to 2000 hours of fan operation.

Blades—Newly designed General Electric "Quiet Blades" shaped and pitched to provide a high-velocity air delivery with practically no sound. Polished aluminum, finished in clear lacquer.

Air delivery—Maximum velocity of 1700 feet per minute delivered by this fan surpasses in cooling power the velocity of conventional fans with larger blades. Volume, 1200 cubic feet per minute.

Guard—Sturdy steel-wire construction with welded joints. Fan may be carried by guard.

Switch—Easy-throw lever switch in the base provides three operating speeds; low, medium and high.

Base—Pressed steel with felt pad to protect furniture from scratches, and with holes to permit mounting on wall or on hanger stud of G-E Elexit Fan Hanger.

Adjustment—Fan can be tilted up or down.

Oscillating mechanism—Turning the knurled compression stud permits the fan to oscillate through an arc of 80 degrees. The friction clutch allows the blades to continue to deliver a breeze should the oscillation of the fan be stopped by some object. This mechanism is packed in high-grade grease which requires renewal only once a season.

Cord—8-ft. length of Underwriters' Laboratories approved and labeled cord, equipped with a G-E Junior All-rubber Attaching Plug.

Finish—Black enamel with polished aluminum blades and chrome trimmings.

Approximate operating cost, 3/5 cent per hour. Overall dimensions of shipping cases, 17¾ in. x 13¾ in. x 18¾ in. Net weight, 16 pounds.

LARGE OSCILLATING FOUR-BLADE QUIET FANS

Cat. No.	Voltage	Cycles	Watts	Loudness Units	List Price
Alternating Current					
*49X496	110	25	68	47	$29.50
49X492	110	50	80	44	28.00
49X494	220	50	80	44	29.50
49X491	110/120	60	85	47	28.00
49X495	220	60	85	47	29.50
Direct Current					
49X493	105/120	..	65	47	$31.50
49X497	210/240	..	65	47	33.00

*Furnished in series-commutator type only.

5

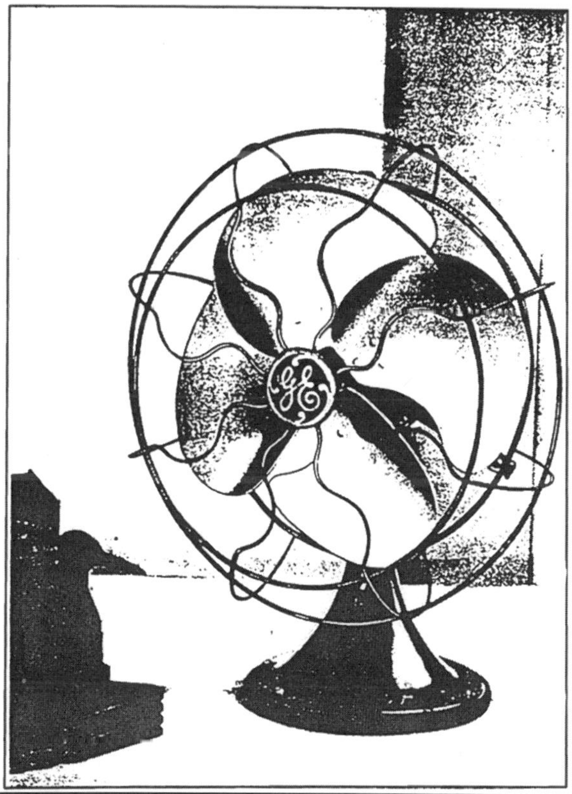

An illustration of a General Electric fan from a 1936 catalog. This 16" fan has newly designed quiet blades that are quieter than standard traditional windmill blades. These fans did not incorporate the ring around the motor and were trimmed in highly polished chrome blades and screws. The black gloss finish and chrome blades made for a dynamic contrast. General Electric was noted for its cuting edge design changes in the industry.

Left: ***GE***
Type: 49X491
Blades: 16"
Date: 1936

Right: ***GE***
Type: 55X165B
Blades: 9"
Date: 1936

On larger quiet blade fans General Electric used four blades instead of three. This fan has a modified traditional S wire guard attached to the motor by three struts. The oscillator ring around the motor has been removed giving the fan a much more modern and streamlined look. The base has a three-speed slide switch that has been made much smaller than those found on earlier fans. This fan is finished in highly gloss black but some examples are found in hunter green.

Grade I – $130.00
Grade II – $98.00
Grade III – $65.00
Grade IV – $45.00
Grade V – $26.00
Grade VI – $7.00

A small fan designed for residential home use. The blades are highly polished aluminum quiet blades with a streamlined guard. The motor is surrounded by an oscillator ring as in previously designed fans. It seemed that GE continued to use the ring around the motor on their 9" fans until almost 1950. This may have been to use up old stock or just not wanting to change the basic design. This fan is finished in high gloss black.

Grade I – $90.00
Grade II – $68.00
Grade III – $45.00
Grade IV – $32.00
Grade V – $18.00
Grade VI – $5.00

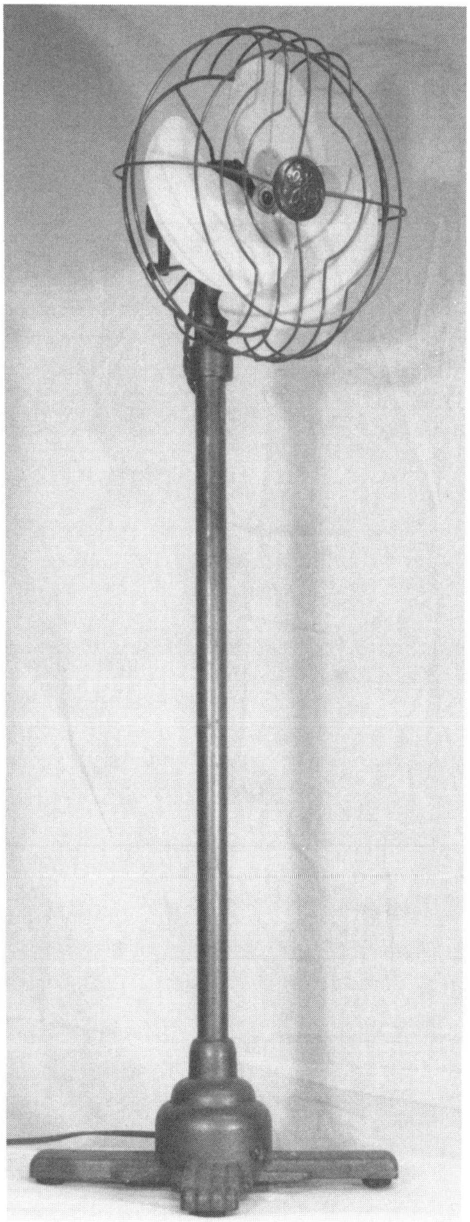

General Electric Pedestal Fan
Type: 49X714
Blades: 12"
Date: 1936

During the thirties, General Electric began manufacturing large numbers of pedestal fans. These fans were marketed for commercial and residential use. These fans were in direct competition with Emerson Seabreeze pedestal fans being manufactured at the same period. This fan incorporates the newly designed overlapping quiet blades and guard. The fan has one speed and is finished in a light metallic brown color. A one-speed toggle switch is located in the base. The base is made of cast iron in a cross configuration and is not very stable. These fans have a tendency to fall over if disturbed. The guard is held to the motor by two machine screws and is not very secure. General Electric improved this design in the late thirties with their Vortalex pedestal fans. Many of these pedestal fans were manufactured and they are quite common.

Grade I –	$150.00	Grade IV –	$53.00
Grade II –	$112.00	Grade V –	$30.00
Grade III –	$75.00	Grade VI –	$8.00

All during the later thirties, GE produced many variations using their older designs as well as introducing new designs.

Left: ***GE***
Type: 75X423
Blades: 12"
Date: c.1938

An industrial grade fan with a newly designed cross guard. This fan is a variation of the fans made in the twenties. The finish is hunter green with a brass and green badge. There is a three-speed slide switch in the base and it will oscillate.

Grade I – $90.00
Grade II – $68.00
Grade III – $45.00
Grade IV – $32.00
Grade V – $18.00
Grade VI – $5.00

Right: ***GE***
Type: 49X719
Blades: 10"
Date: c.1939

This fan has several stylized features such as a deco style octagonal guard, newly designed blades, and an octagonal base. The motor has no oscillator ring and continually oscillates. The base and motor are painted in a wrinkle finish. The base has a one-speed rotary switch. This was a very inexpensive fan that had more looks than function. This style of fan was marketed after World War II as a lower priced fan.

Grade I – $50.00
Grade II – $38.00
Grade III – $25.00
Grade IV – $18.00
Grade V – $10.00
Grade VI – $3.00

GE
Type: unknown
Blades: 10"
Date: 1941

In 1941 General Electric introduced a small modern home use fan. This fan was the first to use the newly designed base with a large strap of sheet steel holding the motor to the base. This design was used on several later fans made by General Electric following World War II. The base is made of brown Bakelite and has "General Electric" embossed across the front. The body of the fan is dark brown with a matching steel guard. The blades are a light brown and are made of one sheet of formed steel. A one-speed rotary switch is located on the top of the bullet (the one in the picture has been replaced) and the fan will oscillate. This was a very inexpensive and poorly constructed fan.

Grade I –	$30.00	Grade IV –	$10.00
Grade II –	$23.00	Grade V –	$6.00
Grade III –	$15.00	Grade VI –	$2.00

Following World War II the fan market was going full speed because air conditioning had not yet become inexpensive enough for home use. General Electric made several grades of fans after the war. The two pictured above are of a lower grade, priced for office or industrial use.

Left: **GE**
Type: FM12541
Blades: 12"
Date: c. 1950

This fan is a late example of a standard small blade fan for commercial or industrial use. The guard is cross type in design, but has been redesigned with no separate struts attaching it to the motor. The guard is held to the motor by wire struts; this was a cost cutting measure. The fan has a two-speed toggle switch in the base and will oscillate. The finish is a metallic sea blue-green and very attractive. The badge is red and silver.

Grade I –	$50.00	Grade IV –	$18.00
Grade II –	$38.00	Grade V –	$10.00
Grade III –	$25.00	Grade VI –	$3.00

Right: **GE**
Type: unknown
Blades: 16"
Date: 1947

This was a very large inexpensive of fan made for commercial use. The blades are made of sheet aluminum and will become out of balance if disturbed. The guard is the cross style with a large black and silver GE badge. The base looks small and not well designed for the 16" blades. A three-speed rotary switch is located in the base. The fan has an oscillator and is finished in gloss black.

Grade I –	$60.00	Grade IV –	$21.00
Grade II –	$45.00	Grade V –	$12.00
Grade III –	$30.00	Grade VI –	$3.00

2,253,066
FAN
Millard F. Dowell, Bridgeport, Conn., assignor to General Electric Company, a corporation of New York
Application December 29, 1938, Serial No. 248,275
8 Claims. (Cl. 170—159)

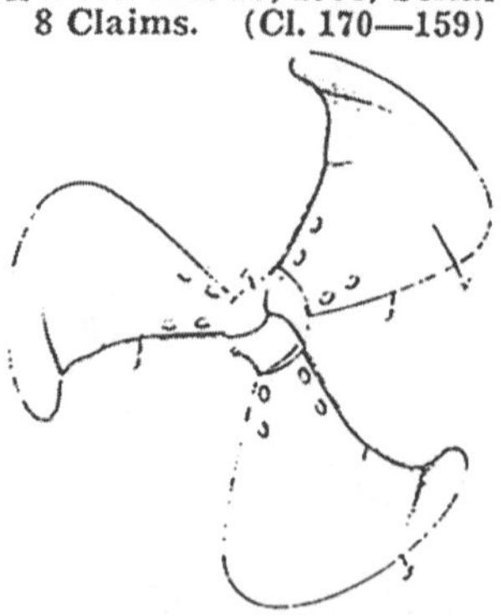

1. In a free air fan, a blade having a trailing tip extending circumferentially from the periphery of the main body of the blade, said tip having a backward curvature as regards the direction of air flow perpendicular to the blade surface, and the periphery of the main body of the blades having a backwardly curved portion increasing in radial width toward and merging into the tip.

In 1941 General Electric patented the Vortalex trailing tip fan blade. This blade became a standard feature on their upscale fans during and after World War II. This illustration is from a 1941 United States patent book.

2,600,325
ENCLOSED OSCILLATING FAN
George J. Reisch, Stratford, Conn., assignor to General Electric Company, a corporation of New York
Application November 13, 1948, Serial No. 59,847
9 Claims. (Cl. 230—256)

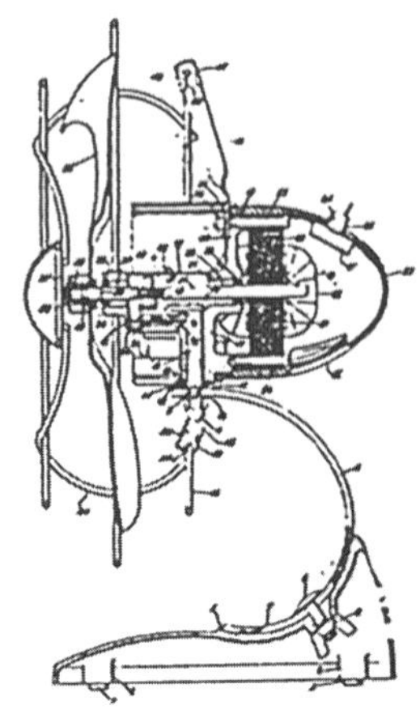

In 1947 General Electric patented a front enclosed oscillator for their small home use fans. This oscillator was met with modest success and discontinued in the fifties.

Following World War II General Electric introduced more Vortalex fans. These new fans had a trailing tip Vortalex blade for home and office use. These fans represented their top of the line fans and were quite expensive. Like Parker blades on Emerson fans, the Vortalex soon became a hallmark of post World War II GE fans. The first Vortalex fans came out about 1941, and were used until about 1950. These fans were very popular and big sellers for the company.

Top Left: ***GE***
Type: 49X642
Blades: 9"
Date: c.1946

This fan has newly designed blades and guard. The motor sits in an oscillator ring attached to a stamped steel base with a one-speed toggle switch. The guard on the 9" fans is rather flimsy because of very thin steel wire used. The badge is sheet aluminum painted with the GE logo and name.

Grade I – $90.00
Grade II – $68.00
Grade III – $45.00
Grade IV – $32.00
Grade V – $18.00
Grade VI – $5.00

Top Right: ***GE***
Type: 49X943B
Blades: 9"
Date: c.1946

This fan is in ivory instead of gloss black. These are not as common as the black examples and are more collectible.

Grade I – $110.00
Grade II – $83.00
Grade III – $55.00
Grade IV – $39.00
Grade V – $23.00
Grade VI – $6.00

Bottom Left: ***GE***
Type: 78X593
Blades: 12"
Date: c.1946

The guard on this fan is made of higher gauge steel and holds up better than the 9" and 10" fans. The blades are highly polished aluminum and are striking with the gloss black finish of the body. A two-speed toggle switch is located in the base, and the motor has a handle for easy transport.

Grade I – $110.00
Grade II – $83.00
Grade III – $55.00
Grade IV – $39.00
Grade V – $23.00
Grade VI – $6.00

Bottom Right: ***GE***
Type: 78X805
Blades: 10"
Date: c.1946

A two-speed oscillating fan with highly polished aluminum Vortalex blades.

Grade I – $90.00
Grade II – $68.00
Grade III – $45.00
Grade IV – $32.00
Grade V – $18.00
Grade VI – $5.00

In the later forties, GE made several cosmetic changes on their top of the line Vortalex fans. These fans were finished in a dark metallic brown with light brown blades. The guard had been changed by leaving off one of the cross wires. These fans were manufactured for several years and were sold in large quantities.

Left: ***GE***
Type: FM10V1
Blades: 10"
Date: c.1947

This fan is the top of the line GE fan for home use with a two-speed toggle switch in the base. This fan will oscillate.

Grade I –	$50.00
Grade II –	$38.00
Grade III –	$25.00
Grade IV –	$18.00
Grade V –	$10.00
Grade VI –	$3.00

Center: ***GE***
Type: FM9V1
Blades: 9"
Date: c.1947

This was one of the very last fans to use an oscillator ring around the motor which looks out of date on this fan. It has a one-speed toggle switch in the base and is finished in light brown.

Grade I –	$70.00
Grade II –	$53.00
Grade III –	$35.00
Grade IV –	$25.00
Grade V –	$14.00
Grade VI –	$3.00

Right: ***GE***
Type: FM12V1
Blades: 12"
Date: c.1947

Most 12" and 16" fans were used for commercial use and this is an example. This 16" fan is massive with deep pitched blades and large guard. One problem with these fans is that they tended to become out of balance if the blades were disturbed in any way. This was a big drawback to the Vortalex blade.

Grade I –	$70.00
Grade II –	$53.00
Grade III –	$35.00
Grade IV –	$25.00
Grade V –	$14.00
Grade VI –	$3.00

Left: ***GE***
Type: F125107
Blades: 10″
Date: 1947

Right: ***GE***
Type: FM10V41
Blades: 10″
Date: c.1950

In the late 1940s GE redesigned their desk fan line with several cost saving measures. The blades were made of a single sheet of formed steel and the guard had larger gauge steel wire finished in highly polished chrome. The motor was covered by a hard plastic shell having the oscillator placed in the front of the motor. The motor was held to the base by a strap of formed steel finished in highly polished chrome. The base was formed steel with "General Electric" stamped over the front. This design was very modern, but not very efficient. The fan had to be turned upside down to adjust the tilt of the fan by using a wing nut underneath the fan. This wing nut could not be adjusted while the fan was running as in older models. The newly designed oscillator was a remake of the 1924 version, and not very good. General Electric was never known for their oscillators and this was no exception.

Grade I – $30.00
Grade II – $23.00
Grade III – $15.00
Grade IV – $10.00
Grade V – $6.00
Grade VI – $2.00

A late example of a Vortalex fan in a sea green metallic finish. These fans were made primarily for commercial or industrial use. The design is exactly the same as the FM10V1, but in a different finish. By the early fifties, the trailing tip Vortalex fans were taken out of production.

Grade I – $40.00
Grade II – $30.00
Grade III – $20.00
Grade IV – $14.00
Grade V – $8.00
Grade VI – $2.00

Westinghouse Electric

Westinghouse Electric was founded by one of the great industrialists of the nineteenth century, George Westinghouse. Westinghouse made his fortune following the Civil War by perfecting an air brake used on railroad cars and had the resources to manufacture high quality fans among other things.

Westinghouse had many very talented people working for him such as Nikola Telsa who understood the importance of electrical power. Telsa was an innovator in alternating current and helped win Westinghouse early electrical related contracts. In 1889, Westinghouse came out with their first desk fan. Westinghouse experimented with several designs in the 1890s and by the turn of the century was a major manufacturer of desk fans.

In 1905, Westinghouse perfected an alternating current fan motor with an internal starting switch. This motor is commonly called a tank motor duc to its size and configuration. These fans were manufactured in great numbers and were very successful.

In 1909, Westinghouse marketed its first vane oscillating fan in 12" and 16" sizes. These fans were very easy to use and fairly inexpensive to manufacture. In 1912, they introduced a mechanical oscillator that was far superior to Emerson and General Electric. This oscillator was used all through the teens, twenties, and early thirties.

1912 was also an important year because the tank motor was discontinued for the stamped steel design. These new fans were much lighter and easier to move than older cast-iron fans.

In 1925, a new man-made plastic blade was introduced. This blade added less total weight to the fan and was much easier to manufacture and balance. This plastic fan blade was far ahead of the competitors in design.

Westinghouse was slow to change its fan design following the Depression. In 1937, they introduced their Power-Aire line with newly designed overlapping blades. This fan was met with minimal success and redesigned again in 1941. The new 1941 design was much improved over the earlier model and sold well. It became a prototype of Westinghouse fans following World War II.

Westinghouse continued to produce high quality fans until the early sixties. They experimented with new designs and colors, but never matched Emerson or GE.

Westinghouse
Type: W-134058A Westinghouse Telsa
Blades: 12"
Date: 1898

Nikola Tesla was a brilliant electrical engineer who worked for Westinghouse in the 1890s. He developed the early alternating current induction motor and was almost solely responsible for the development of alternating current. While Edison preferred direct current, Telsa with the help of Westinghouse, made alternating current the preferred energy source. This fan has Tesla's early induction motor. The motor housing is heavy cast iron. The base is ribbed and made of cast iron. There is a two-speed slide switch in the base. The blades are made of heavy sheet brass and are very attrative. The guard is brass in the S style and very well constructed.

Grade I –	$6,000.00	Grade IV –	$2,100.00
Grade II –	$4,500.00	Grade V –	$1,200.00
Grade III –	$3,000.00	Grade VI –	$300.00

Left: ***Westinghouse***
Type: 28117A
Blades: 12"
Date: 1905

This is the only pancake fan manufactured by Westinghouse and features very heavy cast-iron castings. The guard is made of brass wire with a large cast-brass badge. The badge has a map of the world with the name "Westinghouse" across the front. The blades are made of heavy sheet brass in a large windmill configuration and are attached to a cast-brass hub. This pancake motor has large front and back air vents with a single bearing configuration. A massive trunnion holds the motor to the base. A two-speed switch is located in the base. This fan was not produced in great numbers and was soon replaced by the tank motor fans.

Grade I –	$5,000.00	Grade IV –	$1,750.00
Grade II –	$3,750.00	Grade V –	$1,000.00
Grade III –	$2,500.00	Grade VI –	$250.00

WESTINGHOUSE FAN MOTORS FOR ALTERNATING CIRCUITS.

Swivel and Trunnion Type.

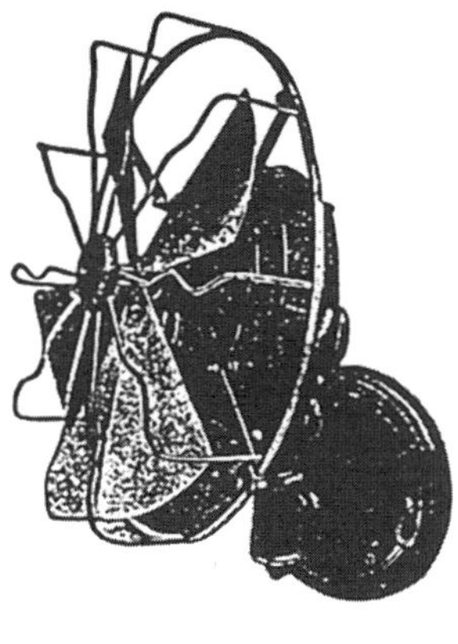

Bracket Fan.

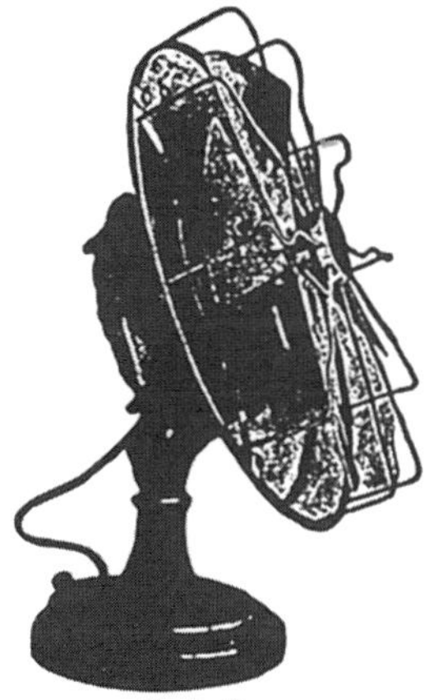

Desk Fan.

The mechanical parts are of a graceful outline and present a pleasing appearance. The iron parts are japanned, and the fan blades, guards and oil-cups are of polished brass.

Two speeds are obtained by the use of a controlling switch mounted in the base of the motor. This switch is of simple and durable construction, and cannot get out of order.

The high speed gives force and volume enough to insure ventilation to a large room, while the low speeds are best suited for desk fans and consume considerably less energy.

The bearings are liberally designed, and are lubricated by means of oil-cups containing felt wicks. The oil-cups are filled before the motor is tested, and the bearings require no attention after the motor is installed.

The motors are made in both the bracket and desk types. In both styles the motor is hung upon a trunnion, the trunnion fork being mounted on a swivel. The fan motor and guard are nearly balanced on the trunnion, and thumb-screws at either side serve to hold the fan motor from tipping. The swivel turns readily, but is adjusted by means of a spring to prevent the fan from turning of itself.

The bracket type fan may be adapted for use as a desk type fan by removing the swivel elbow, or adapter, and inserting the stem of the trunnion fork into the base.

Deterioration of the insulation by heat is impossible, owing to the circulation of air through the windings of the fields and armature. This is made possible by openings pierced in the motor frame.

All parts carrying an electric current are well insulated, and it is impossible for a person touching the fan or motor to receive a shock.

Code word	Style No.	Type	Size	Voltage	Alternations	Speed High	Speed Slow	List price each
Instantany	28,117	Desk	12	100-120	7,200	1,550 R. P. M.	1,000 R. P. M.	$24.00
Instanter	28,120	Desk	12	100-120	16,000	1,450 R. P. M.	1,200 R. P. M.	24.00
Instantly	28,738	Bracket	12	100-120	7,200	1,550 R. P. M.	1,000 R. P. M.	26.40
Instar	28,741	Bracket	12	100-120	16,000	1,450 R. P. M.	1,200 R. P. M.	26.40
Instatement	28,119	Desk	16	100-110	7,200	1,550 R. P. M.	1,000 R. P. M.	31.20
Instillator	28,122	Desk	16	100-110	16,000	1,450 R. P. M.	1,200 R. P. M.	31.20
Instiller	28,740	Bracket	16	100-110	7,200	1,550 R. P. M.	1,000 R. P. M.	33.60
Instilment	28,743	Bracket	16	100-110	16,000	1,450 R. P. M.	1,200 R. P. M.	33.60

Power Required.—12-inch Fan—Maximum speed, 65 watts; minimum speed, 56 watts. 16-inch Fan—Maximum speed, 103 watts; minimum speed, 87 watts.

An illustration from a 1904 Illinois Elcctric Company catalog showing a Westinghouse pancake fan. This was the only pancake style fan manufactured by Westinghouse and is very rare. This fan also came in a bracket fan as illustrated.

Left: ***Westinghouse***
Type: Tank Motor
Blades: 12"
Date: 1907

An example of an early tank motor fan. The term tank motor comes from the large size of the motor. The motor has very heavy castings with large vent holes. These early motors had internal starting switches and were produced in great numbers. The guard is made of brass wires in an S style. The guard badge is a strip of cut out brass with "Westinghouse" across it. The early tank motors have the brass letters cut out across the banner on the badge, the later ones were just stenciled.

Grade I –	$200.00	Grade IV –	$70.00
Grade II –	$150.00	Grade V –	$40.00
Grade III –	$100.00	Grade VI –	$10.00

Right: ***Westinghouse***
Type: Tank Motor
Blades: 12"
Date: 1912

An example of a later tank motor. The strip across the front of the badge is painted with the name "Westinghouse." These fans were manufactured in great quantities and were very successful for the company.

Grade I –	$200.00	Grade IV –	$70.00
Grade II –	$150.00	Grade V –	$40.00
Grade III –	$100.00	Grade VI –	$10.00

Around 1910, most of the major fan manufactures were in the process of designing an oscillator that would be cheap to manufacture, easy to use, and efficient. Charles Eck had developed a mechanical oscillator in about 1903, but it was not until around 1910 that most other manufacturers began producing oscillators. Westinghouse was unusual in the fact that they developed vane and mechanical oscillators at the same time. Emerson and General Electric never marketed a vane oscillator and preferred mechanical oscillating devices.

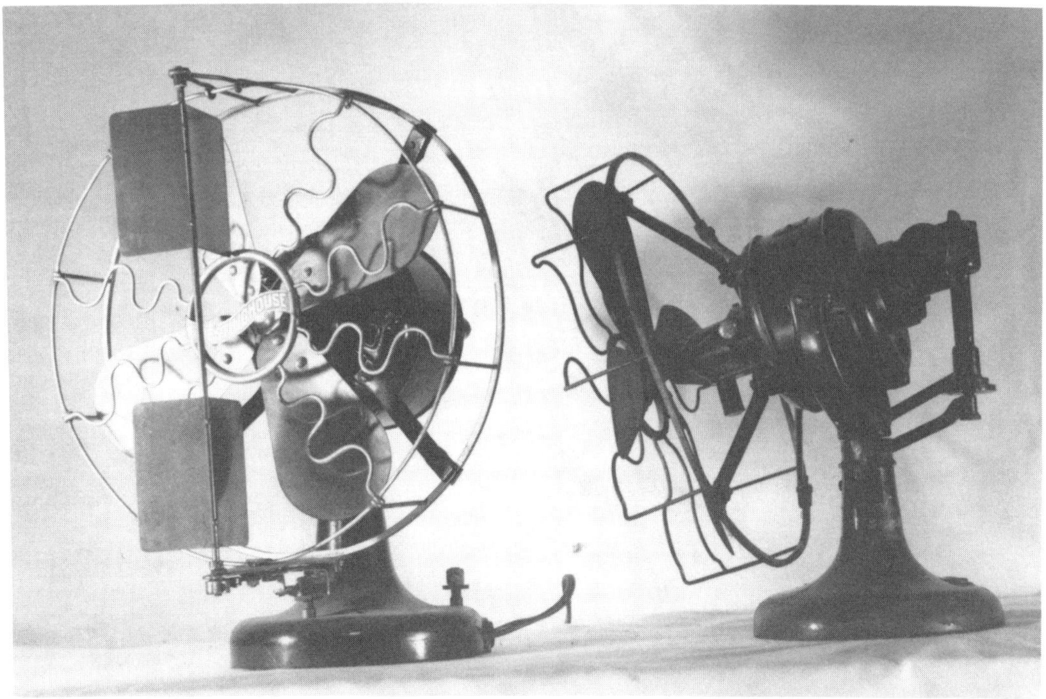

Left: ***Westinghouse***
Type: Vane Oscillator
Blades: 12"
Date: 1911

Right: ***Westinghouse***
Type: Sidewinder
Blades: 12"
Date: 1912

Westinghouse mass produced this fan and it is the most common vane oscillator. This device moved the fan from side to side by the use of a pair of vanes located at the front of the guard. A spring device at the base of the vane tripped the vane allowing the fan to rotate back and forth. When the vane was in proper alignment and on a level surface, it worked quite well. Knobs used as stops located at the base of the fan would regulate the width of the arc of oscillation. The base is heavy cast-iron with a two-speed switch. The tank motor has an internal starting switch with large air vents. The motor tag is held to the motor housing by small brass screws; this was unusual since most motor badges were held by rivets. The guard is held to the motor by large cast-iron struts.

Grade I – $1,100.00
Grade II – $830.00
Grade III – $550.00
Grade IV – $390.00
Grade V – $230.00
Grade VI – $60.00

An example of a sidewinder with two levers protruding from the base to the motor. This fan is a direct current example, but they also came in alternating current. The motor is very sleek and modern looking. A two-speed switch is located in the base. The blades are made of heavy sheet brass with a brass guard.

Grade I – $3,000.00
Grade II – $2,200.00
Grade III – $1,500.00
Grade IV – $1,050.00
Grade V – $600.00
Grade VI – $150.00

Left: ***Westinghouse***
Type: 149228 "Sidewinder"
Blades: 12"
Date: 1910

An example of an alternating current sidewinder oscillator. The motor housing is made of cast iron with large vent holes and has an internal starting switch. The base is cast iron with a two-speed switch. The motor sits on a trunnion and pivots back and fourth by use of two levers attached to base from the oscillator wheel. The blades are made of heavy sheet brass, as is the guard. Very rare.

Grade I – $4,000.00
Grade II – $2,200.00
Grade III – $2,000.00
Grade IV – $1,400.00
Grade V – $800.00
Grade VI – $200.00

Right: ***Westinghouse***
Type: 133598 "Sidewinder"
Blades: 12"
Date: 1910

This fan was manufactured by General Electric, but has a Westinghouse guard and motor badge. There seems to have been some borrowing between General Electric and the other major manufacturers. Emerson made a direct current pancake fan using a GE motor and guard, but having Parker blades and an Emerson motor badge. This fan would be an example of such borrowing. The fan for all intensive purposes is a GE with a different guard and is very collectible.

Grade I – $3,000.00
Grade II – $2,250.00
Grade III – $1,500.00
Grade IV – $1,050.00
Grade V – $600.00
Grade VI – $150.00

Westinghouse
Type: 60673
Blades: 12"
Date: 1912

A rear view of a direct current non-oscillating fan. The motor is completely enclosed with the exception of exposed brushes at the rear of the fan. The motor is held to the base by a cast-iron trunnion. The base is heavy cast iron with a two-speed slide switch. The blades and guard are made of brass.

Grade I –	$1,000.00	Grade IV –	$350.00
Grade II –	$750.00	Grade V –	$200.00
Grade III –	$500.00	Grade VI –	$50.00

In 1912, Westinghouse introduced a new line of fans with a stamped steel motor housing and base. This was a revolutionary move since most fans of the period were made of cast iron. The stamped steel parts made the fan much lighter and easier to move from place to place, especially when used in the home. This was a selling point used by Westinghouse to market their fans. The new motors were much sleeker and more modern looking than the tank motors had been. That same year they introduced their new improved mechanical oscillator that was completely enclosed and could be engaged by lifting a small knob located at the rear of the fan. The first of these new fans had internal starting switches, but later examples did not.

Left: ***Westinghouse***
Type: 164864B
Blades: 12"
Date: 1914

A residential six-blade oscillator having the newly designed stamped steel features. The base has a three-speed slide switch. The blades are heavy sheet brass attached to a brass hub. The guard is brass in an S style attached to the motor by stamped steel struts. This example has an internal starting switch and is very quiet. As residential oscillators of the period, it was far superior to Emerson or General Electric being lighter and having an easy to use oscillator

Grade I – $500.00
Grade II – $375.00
Grade III – $250.00
Grade IV – $175.00
Grade V – $100.00
Grade VI – $35.00

Right: ***Westinghouse***
Type: W134110
Blades: 8"
Date: 1912

Westinghouse made many fans for Western Electric, which manufactured no fans. This fan is all brass finished in gun blue. A three-speed slide switch is located in the base. The universal motor is very compact and high speed. The guard is very flimsy and made out of thin brass wire. The blades are made of sheet brass in the windmill style and attached to a brass hub.

Grade I – $300.00
Grade II – $250.00
Grade III – $150.00
Grade IV – $106.00
Grade V – $60.00
Grade VI – $15.00

Small fans for residential home use became very popular around 1910. Small fans were less expensive and easier to market as electricity became more available.

Left: ***Westinghouse***
Type: unknown
Blades: 8"
Date: 1912

This is a beautiful all brass fan finished in polished brass. This fan has three speeds and will not oscillate. The blades are windmill type and the guard is quite delicate. This fan has a universal motor.

Grade I – $300.00
Grade II – $250.00
Grade III – $150.00
Grade IV – $106.00
Grade V – $60.00
Grade VI – $15.00

Center: ***Westinghouse***
Type: unknown
Blades: 9"
Date: c.1922

This fan is unusual because it is an oscillator. Emerson and General Electric both produced small 9" oscillating fans, but Westinghouse seems to have not manufactured them in such great numbers. This example has stamped steel rounded blades, an open wire S guard with a three-speed slide switch in the base. Shortly after this fan was manufactured, Westinghouse began to use plastic blades instead of stamped steel.

Grade I – $70.00
Grade II – $53.00
Grade III – $35.00
Grade IV – $25.00
Grade V – $14.00
Grade VI – $3.00

Right: ***Westinghouse***
Type: Whirlwind
Blades: 9"
Date: c.1919

The Whirlwind line was an inexpensive group of fans introduced around World War I and manufactured until the middle twenties. The first group had an open S wire guard similar to the all brass fans manufactured a few years earlier. The blades were made of stamped steel in the windmill style. The motor is an induction type and not universal as on earlier fans. Some early examples have brass blades, but those are rather uncommon. This is a plug-in fan with no switch in the base.

Grade I – $50.00
Grade II – $38.00
Grade III – $25.00
Grade IV – $18.00
Grade V – $10.00
Grade VI – $3.00

Left: ***Westinghouse***
Type: 165366
Blades: 9″
Date: c.1913

An all brass fan for 220 volts. Base and motor housing are finished in gloss black. A three-speed slide switch is located in the base. The fan will not oscillate. Unusual voltage is a unique feature and makes this fan very collectible.

Grade I – $500.00
Grade II – $375.00
Grade III – $250.00
Grade IV – $175.00
Grade V – $100.00
Grade VI – $35.00

Right: ***Westinghouse***
Type: Whirlwind
Blades: 9″
Date: c.1922

A later example of a Whirlwind with a universal motor. Note that the brush caps protrude out directly from the back plate of the motor, this is an unusual design. The base is cast iron and there is no switch. A very high speed fan that blows a lot of air for its small size.

Grade I – $50.00
Grade II – $38.00
Grade III – $25.00
Grade IV – $18.00
Grade V – $10.00
Grade VI – $3.00

During the twenties, Westinghouse produced many quality desk fans for consumer use. They tended to use a design developed around 1916 and made minor adjustments until the Depression. These fans were popular with consumers and many were produced.

Left: ***Westinghouse***
Type: 31575-A
Blades: 12"
Date: c.1925

In 1924, Westinghouse patented a man-made plastic material to be used for their fan blades. This material was lightweight and easy to manufacture. The lightweight blades made them easy to balance and put less strain on the motor. This fan incorporates these new blades. This is an early example because the guard is an open S type used on fans from the early twenties. The base and motor housing are made of stamped steel. A three-speed slide switch is located in the base and this fan will oscillate.

Grade I –	$90.00
Grade II –	$68.00
Grade III –	$45.00
Grade IV –	$32.00
Grade V –	$18.00
Grade VI –	$5.00

Center: ***Westinghouse***
Type: 1648641
Blades: 12"
Date: c.1923

Westinghouse made residential fans until the middle twenties and then seemed to have stoped manufacturing them. This is a late example of a Westinghouse residential fan. The guard is a steel open S wire attached to the motor by four stamped steel struts. The motor housing and base are stamped steel finished in gloss black. This fan has three speeds and will oscillate. By the twenties, residential fans were falling out of favor with consumers.

Grade I –	$150.00
Grade II –	$112.00
Grade III –	$75.00
Grade IV –	$53.00
Grade V –	$30.00
Grade VI –	$8.00

Right: ***Westinghouse***
Type: 211853-E
Blades: 10"
Date: c.1923

This fan incorporates several important features that made Westinghouse fans popular such as being lightweight and easy to use. The fan is made of all stamped steel parts and is easy to move from place to place. The Westinghouse oscillator was easy to use and worked very efficiently. The oscillator had an enclosed gear box and lever system that gave the fan a sleek modern look. Three speeds were a standard feature on all Westinghouse 10" fans at this time.

Grade I –	$60.00
Grade II –	$45.00
Grade III –	$30.00
Grade IV –	$21.00
Grade V –	$12.00
Grade VI –	$3.00

Left: ***Westinghouse***
Type: unknown
Blades: 10"
Date: c.1932

This fan incorporates several interesting design changes from the typical twenties Westinghouse desk fan. The guard is completely redesigned and unusual for a Westinghouse fan. The wire stock on the guard is flat with a hexagonal badge. The base is a modified cone design with Art Deco styling. The motor and blades are very characteristic of earlier Westinghouse fans. This fan will oscillate.

Grade I – $70.00
Grade II – $53.00
Grade III – $35.00
Grade IV – $25.00
Grade V – $14.00
Grade VI – $3.00

Right: ***Westinghouse***
Type: 457680
Blades: 10"
Date: 1924

In 1924, Westinghouse came out with a new color for their standard 10" desk fan. This new lighter color did not hold up as well as the traditional black finish and was soon abandoned. The blades, motor housing, and base are made of stamped steel finished in a light yellow. There is a three-speed slide switch in the base and the fan will oscillate.

Grade I – $50.00
Grade II – $38.00
Grade III – $25.00
Grade IV – $18.00
Grade V – $10.00
Grade VI – $3.00

Left: ***Westinghouse***
Type: 8-SFN-30
Blades: 9"
Date: c.1927

A small direct current fan with a black wrinkle finish. A three-speed switch is located in the base, and the fan will not oscillate. This fan is unusual because it is for direct current and is such a small size.

Grade I – $90.00
Grade II – $68.00
Grade III – $45.00
Grade IV – $32.00
Grade V – $18.00
Grade VI – $5.00

Center: ***Westinghouse***
Type: 214842
Blades: 10"
Date: c.1919

This fan is unusual because the blades are made of sheet brass-plated nickel. Nickel plating was used on several early fans made by Emerson, Peerless, and during the twenties by StarRite. This was an unusual finish to use on the blades. Besides the nickel-plated blades, the guard is held to the motor by three struts instead of four. The motor, base, and switch are all similar to fans produced right after World War I. This is a non-oscillator.

Grade I – $130.00
Grade II – $98.00
Grade III – $65.00
Grade IV – $45.00
Grade V – $26.00
Grade VI – $7.00

Right: ***Westinghouse***
Type: 77572
Blades: 9"
Date: c.1927

By the later twenties Westinghouse seems to have dropped the Whirlwind line of small fans and just produced 9" fans under the Westinghouse banner. This example has a cast-iron base and is one speed. A small slide switch is located on the top of the motor making this an unusual fan.

Grade I – $50.00
Grade II – $38.00
Grade III – $25.00
Grade IV – $18.00
Grade V – $10.00
Grade VI – $3.00

After the Depression, Westinghouse changed their guard by abandoning the traditional S wire style for a more modern spoke guard that was easier to manufacture and had a more streamlined appearance. These were mass produced fan made for the middle range market. The Power-Aire fans were designed for the more affluent consumer.

Left: ***Westinghouse***
Type: 420550-A Whirlwind
Blades: 9″
Date: c.1922

This is a very late example of a Whirlwind fan. The base is cast iron with a one-speed slide switch. The universal motor is very high speed and has rear brush caps.

Grade I	$50.00
Grade II	$38.00
Grade III	$25.00
Grade IV	$18.00
Grade V	$10.00
Grade VI	$3.00

Center: ***Westinghouse***
Type: 803008-A
Blades: 10″
Date: c.1933

An example of an early Depression fan. The guard has been modified, but the blades, motor, and base are very typical of the twenties. This fan has three speeds and will oscillate, non-oscillators were also manufactured in this style.

Grade I	$50.00
Grade II	$38.00
Grade III	$25.00
Grade IV	$18.00
Grade V	$10.00
Grade VI	$3.00

Right: ***Westinghouse***
Type: 868872
Blades: 12″
Date: c.1940

This fan has modified plastic blades that almost look like newer Parker blades made by Emerson after 1940. This fan would have been made for commercial or home use and was a lower grade fan. It has three speeds and will oscillate.

Grade I	$50.00
Grade II	$38.00
Grade III	$25.00
Grade IV	$18.00
Grade V	$10.00
Grade VI	$3.00

In 1937, Westinghouse introduced a new line of fans for the more affluent consumer called Power-Aire. These fans had newly designed blades, guard, and base. The motor was a standard twenties motor covered with an ornate shroud that made an interesting configuration. The Power-Aire line was manufactured in some form until the middle 1950s and was a very successful product for the company.

Left: ***Westinghouse***
Type: 922282 Power-Aire
Blades: 10"
Date: 1937 – 1940

A new concept fan with modern streamlined designing. The blades are deep pitched plastic and overlapping. The guard is made of formed wire with a new style badge. The base is made ot cast iron in an Art Deco style. A one-speed toggle switch is located in the base. This fan seems to be over designed with massive features. The base looks awkward and too large. The blades protrude out from the motor making the fan look massive. The fan is finished in a dark brown metallic paint.

Grade I – $130.00
Grade II – $98.00
Grade III – $65.00
Grade IV – $45.00
Grade V – $26.00
Grade VI – $7.00

Center: ***Westinghouse***
Type: 12PA2 Power-Aire
Blades: 12"
Date: 1941 – 1949

In 1940, Westinghouse completely redesigned their Power-Aire line and reintroduced it in 1941 right before World War II. The 1937 model was entirely overhauled for this new Power-Aire. The shape of the blades were made to look more sleek and were held to the hub by a large chrome-plated button. The guard was modified and the cross badge logo discontinued. The motor was completely redesigned using a General Electric type oscillator abandoning the 1912 Westinghouse oscillator. This was unfortunate because the Westinghouse oscillator was a far superior design to the General Electric. The base was made less massive and has a chrome plate in the front. The fan was very quiet and efficient.

Grade I – $70.00
Grade II – $53.00
Grade III – $35.00
Grade IV – $25.00
Grade V – $14.00
Grade VI – $3.00

Right: ***Westinghouse***
Type: 10-A-3 Power-Aire
Blades: 10"
Date: 1941 – 1949

This fan was designed for home or office use and was very quiet and efficient. This fan was in direct competition with General Electric's Vortalex trailing trip fans of the same period.

Grade I – $50.00
Grade II – $38.00
Grade III – $25.00
Grade IV – $18.00
Grade V – $10.00
Grade VI – $3.00

In the fifties, Westinghouse modified their Power-Aire line of fans again.

Left: ***Westinghouse***
Type: 1381675
Blades: 10″
Date: c.1958

This fan has a cast-aluminum base and motor housing. The blades are stamped steel with a chrome finish. The guard is made of formed wire with a foil badge across the front. The motor is trimmed with silver ribbing and has a one-speed rotary switch in the rear. The oscillator must be changed manually from the base of the oscillator wheel. The fan is cream color with silver trim, and is also found in light blue. There are no oil holes to lubricate the motor making it a throw away fan and a predecessor of the plastic fans of today.

Grade I – $30.00
Grade II – $23.00
Grade III – $15.00
Grade IV – $10.00
Grade V – $6.00
Grade VI – $2.00

Right: ***Westinghouse***
Type: 165D3
Blades: 16″
Date: c.1955

A late Power-Aire example. The base and motor housing are made of cast aluminum. The plastic blades are rounded and modern in design. The guard is typical of the 1940 Power-Aire, but is chrome plated. The oscillator is GE in type and has three speeds. The fan is finished in light mauve.

Grade I – $50.00
Grade II – $38.00
Grade III – $25.00
Grade IV – $18.00
Grade V – $10.00
Grade VI – $3.00

Miscellaneous

Menominee

Menominee fans were manufactured between the late 1890s and World War I. They produced a very interesting line of early fans that are very collectible. The company was sold around 1919 and the former owner moved to Cairo, Illinois, and started the Ideal Electric Company. The Ideal Electric Company produced fans for several years which were very similar to their Menominee predecessors.

Left: ***Menominee***
Type: unknown
Blades: 10″
Date: c.1908

An early Menominee fan with a cast-iron universal ball motor. The guard is made of brass formed wire in the S style. The center badge is a cast-brass open ring. The guard is held to the motor by cast-brass struts. The blades are made of heavy sheet brass in the windmill style. An early slide switch is located in the base and the fan is finished in gloss black.

Grade I –	$1,100.00
Grade II –	$850.00
Grade III –	$550.00
Grade IV –	$390.00
Grade V –	$230.00
Grade VI –	$60.00

Right: ***Menominee***
Type: unknown
Blades: 9″
Date: c.1912

This is an example of a tab foot non-oscillator desk fan. The base and motor housing are heavy cast iron and the guard is made of formed brass wire in the S style. The guard is held to the motor by four cast-brass struts. The blade is made of thick sheet brass in the windmill style. The universal motor has three speeds.

Grade I –	$700.00
Grade II –	$525.00
Grade III –	$350.00
Grade IV –	$245.00
Grade V –	$140.00
Grade VI –	$35.00

Right: ***Menominee***
Type: unknown
Blades: 16″
Date: c.1912

This fan is believed to be an industrial grade fan. It may have been made especially for some individual need. It has a characteristic Menominee guard and badge. The blades are very unusual looking and seem unfinished. The universal motor is set in a strap steel trunnion attached to a massive cast-iron base. The speed of the fan can be changed by moving the brushes up or down. This is an unusual method of speed regulation, but was not unknown. Hamilton Beach and Crocker/Curtis used the same method of speed regulation on some of their small fans. The fan blows a large amount of air and is quiet. The original owner may have needed these features for his office or plant. A very ugly fan with no artistically redeeming features.

Grade I – $700.00
Grade II – $525.00
Grade III – $350.00
Grade IV – $245.00
Grade V – $140.00
Grade VI – $35.00

Left: ***Menominee***
Type: unknown
Blades: 12"
Date: c.1916

Right: ***Menominee***
Type: unknown
Blades: 12"
Date: c.1912

Menominee manufactured residential fans as part of their regular line of products. This example has a large pancake motor with an oscillator located in the rear. The castings are massive and very heavy. The motor is surrounded by a large cast-iron ring in which the motor pivots for oscillation. This is similar to a design used by General Electric. The cone base has a four-speed slide switch which is another unusual feature. The guard is made of formed brass wire in the S style. The guard is held to the motor by cast-brass struts, a hallmark of Menominee fans. The blades are made of thick sheet brass in the windmill style.

Grade I – $1,100.00
Grade II $830.00
Grade III – $550.00
Grade IV – $390.00
Grade V – $230.00
Grade VI – $60.00

This fan has a tab-footed base with a four-speed slide switch. The pancake motor is made of heavy cast iron with an oscillator attached to the rear bearing. The guard is not original, but a later example. The original guard had an open ring badge. The blades are made of heavy sheet brass and are in the windmill style.

Grade I – $800.00
Grade II – $600.00
Grade III – $400.00
Grade IV – $280.00
Grade V – $160.00
Grade VI – $40.00

Left: ***Menominee***
Type: unknown
Blades: 9"
Date: c.1917

Some of the most collectible fans made are Menominee 9" oscillators. This fan also came in with a tab-footed base and is rare. The motor and base are made of cast iron. The motor pivots within an oscillator ring. The oscillating mechanism protrudes several inches from the rear of the motor, giving the fan a strange configuration. The guard is made of formed brass wire with a cast-brass open ring badge. The blades are made of heavy sheet brass in the windmill style. Note the large brass tag attached to the base above the switch. This was a standard feature on most Menominee fans. Menominee tended to put the tag on the base instead of the motor.

Grade I – $900.00
Grade II – $675.00
Grade III – $450.00
Grade IV – $315.00
Grade V – $180.00
Grade VI – $45.00

Center: ***Menominee***
Type: unknown
Blades: 9"
Date: c.1910

This fan is made of two castings put together in a very crude configuration. The guard and blades are steel and very cheaply constructed. The universal motor has no on/off switch and is a plug in fan.

Grade I – $150.00
Grade II – $113.00
Grade III – $75.00
Grade IV – $53.00
Grade V – $30.00
Grade VI – $8.00

Right: ***Menominee***
Type: unknown
Blades: 10"
Date: c.1918

This is one of the last models made by Menominee. It has modern type blades and guard, with a universal motor that is attached to a cone base. This is very similar to Ideal fans made in Cairo, Illinois, in about 1920.

Grade I – $120.00
Grade II – $90.00
Grade III – $60.00
Grade IV – $42.00
Grade V – $24.00
Grade VI – $6.00

Left: ***Menominee***
Type: 108
Blades: 12"
Date: c.1919

A very late example of a Menominee desk fan. The guard is made of formed steel wire in the S style. The guard is spot welded and attached to the motor by steel struts. The base and motor housing are cast iron. There is a three-speed slide switch in the base and this example is a non-oscillator.

Grade I – $150.00
Grade II – $113.00
Grade III – $125.00
Grade IV – $53.00
Grade V – $30.00
Grade VI – $8.00

Right: ***Ideal***
Type: Ideal
Blades: 10"
Date: c.1921

Ideal fans were made by the founder of Menominee when he moved to Cairo, Illinois. This fan has many features found on late model Menominee fans. Note that the information badge is located on the base of the fan, a hallmark of Menominee fans. This fan has a universal motor, two speeds, and is a non-oscillator.

Grade I – $90.00
Grade II – $68.00
Grade III – $45.00
Grade IV – $32.00
Grade V – $18.00
Grade VI – $5.00

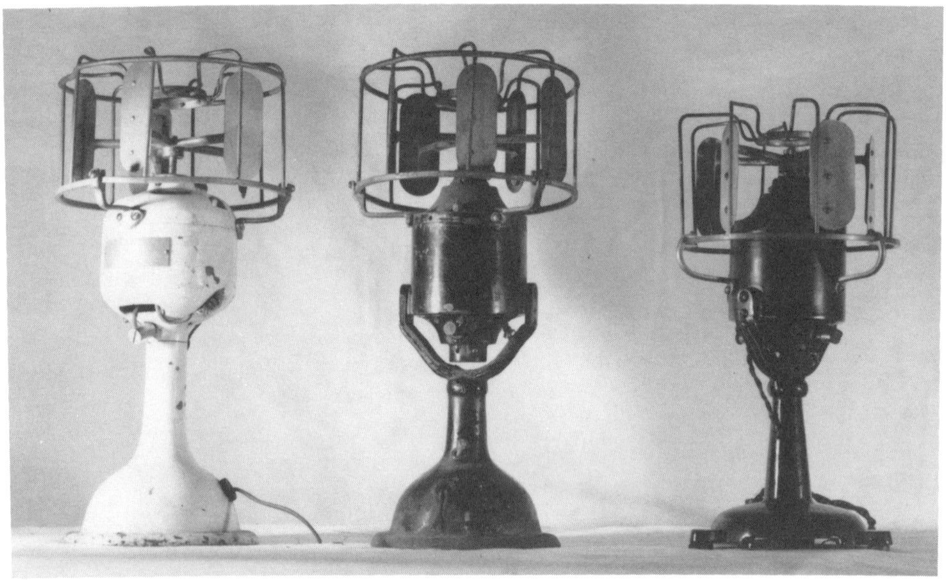

Menominee is best known to collectors for its radial axis, or more commonly called teller fans. These strange configured fans were designed to move air in a 180 degree pattern without disturbing papers on a desk. The blades located on top of the fan rotated at a high speed pulling air from above and below and pushing the air outwards. It seems that these fans met with modest success.

Left: ***Menominee***
Type: Radial-Axis
Date: c.1911

This is the last radial axis model made by Menominee. It has a stamped steel base and motor housing with a three-speed slide switch. The guard is made of formed wire brass with a cast-brass open ring badge. The guard brackets are made of cast brass. The blades are made of aluminum and attached to a cast-brass spider. Aluminum was a very unusual material for such an early fan.

Grade I –	$1,300.00
Grade II –	$975.00
Grade III –	$650.00
Grade IV –	$455.00
Grade V –	$260.00
Grade VI –	$65.00

Center: ***Menominee***
Type Radial-Axis
Date: c.1910

A transitional radial axis fan using a cast-iron base and motor housing attached to a trunnion. The trunnion is used to hold the motor only and will not adjust up or down. The blades are made of sheet aluminum attached to a brass spider. The guard and brackets are brass. This has a three-speed slide switch in the base.

Grade I –	$1,300.00
Grade II –	$975.00
Grade III –	$650.00
Grade IV –	$455.00
Grade V –	$260.00
Grade VI –	$65.00

Right: ***Menominee***
Type: Radial Axis
Date: 1909

The prototype of a radial axis fan. The cast-iron tab-footed base has a three-speed slide switch. The motor is connected to the base by a trunnion. The brass brush caps are exposed and are located at the base of the motor. The blades are made of sheet aluminum connected to a cast-brass spider. The guard is made of brass with a cast-brass open ring badge. The fan is finished in dark matte green.

Grade I –	$2,400.00
Grade II –	$1,800.00
Grade III –	$1,200.00
Grade IV –	$840.00
Grade V –	$480.00
Grade VI –	$120.00

Right: ***R & M***
Type: R136028 "Feather vane"
Blades: 12"
Date: 1912

Robbins & Meyers developed the vane oscillator to its highest level. In 1912 they came out with a vane oscillator that had a device in front of the blades that looked like a feather (missing in this picture). This vane would move back and forth triggered by a mechanism located at the rear of the fan. What separated this device from the Westinghouse vane oscillator is that the motor could be moved up or down without stopping oscillation. By contrast, the Westinghouse fan would not oscillate back and forth if it was not on a level surface. This was a very complex oscillating mechanism and was not manufactured for very long. The large tank motor sits in a C frame that allows the motor to move back and forth. The base is made of cast iron and has a three-speed slide switch. The guard is made of heavy brass wire in the S style. Note how deep the guard is to accommodate the feather vane. The blades are made of very heavy sheet brass with a hole in the middle of the motor shaft for the feather vane. There is gold pin stripping along the motor and base.

Grade I – $8,000.00*
Grade II – $6,000.00*
Grade III – $4,000.00*
Grade IV – $2,800.00*
Grade V – $1,600.00*
Grade VI – $400.00*

*Note that this price reflects a complete, working feather-vane, not a partially complete fan as pictured above.

Right: ***R & M***
Type: 1401
Blades: 12"
Date: 1910

A large tank motor fan on a trunnion with beautiful pin stripping on the motor and base. The motor, trunnion, and base are made of heavy cast iron. The blades and guard are made of brass. The R&M flag is missing in the center ring on the guard. The fan is very quiet when in operation and does not have a internal starting switch which is unusual for a tank motor.

Grade I – $300.00
Grade II – $225.00
Grade III – $150.00
Grade IV – $105.00
Grade V – $60.00
Grade VI – $15.00

Robbins and Meyers made large quantities of fans at their company in Ohio. Their fans were of high quality and attractive. Their "standard" fans were very popular and mass produced. Pictured above are two of the most common standard fans, one being a tank motor, the other a stamped steel version.

Left: ***R & M***
Type: 1125
Blades: 16"
Date: c.1913

This is a direct current fan having all the characteristics of a standard R & M tank motor fan. The base and motor housing is made of heavy cast iron. The blades are made of heavy sheet brass. The blades are rounded at the ends similar to the Westinghouse fans of the period. The guard is made of heavy formed brass wire in the S style. At the center of the guard is a ring with a large cast flag, another hallmark of R & M fans. The guard is held to the motor by cast-iron struts. The base has a three-speed slide switch.

Direct Current Fan	Alternating Current Fan
Grade I – $700.00	Grade I – $200.00
Grade II – $525.00	Grade II – $150.00
Grade III – $350.00	Grade III – $100.00
Grade IV – $245.00	Grade IV – $70.00
Grade V – $140.00	Grade V – $40.00
Grade VI – $35.00	Grade VI – $10.00

Right: ***R & M***
Type: 2506
Blades: 16"
Date: c.1919

A later version of a R & M standard fan for direct current. This fan has a stamped steel motor housing and base. The guard is made of steel wire and spot welded. The blades are heavy sheet brass and have a polished finish. The oscillator used on these fans was rather awkward and rotated the fan very quickly back and forth. This oscillator was not very efficient or easy to use. The direct current examples are very rare while the alternating current examples are more common.

Direct Current Fan	Alternating Current Fan
Grade I – $500.00	Grade I – $150.00
Grade II – $375.00	Grade II – $113.00
Grade III – $250.00	Grade III – $75.00
Grade IV – $175.00	Grade IV – $53.00
Grade V – $100.00	Grade V – $30.00
Grade VI – $35.00	Grade VI – $8.00

During the twenties, Robbins and Meyers made many fans for home and office use. These fans were produced in large numbers and sold well. Many of their early small fans had universal motors that were able to use both alternating and direct current. This may have been done because of the availability of direct current in Ohio in the early years of this century.

Left: ***R & M***
Type: 3500
Blades: 8″
Date: c.1919

A small fan with a very unusual switch in the base. The fan has a slide switch that has "Off, AC, DC" printed above the switch. This was an interesting way of marketing a duel use fan. The base and motor housing are made of cast iron and this is a non-oscillating fan.

Grade I – $70.00
Grade II – $53.00
Grade III – $35.00
Grade IV – $25.00
Grade V – $14.00
Grade VI – $3.00

Right: ***R & M***
Type: 2510
Blades: 10″
Date: c.1920

What makes this fan interesting is that is has five brass blades. Five is a very unusual number for fan blades since there are special problems with balancing a blade with five wings. The blades are made of sheet brass and polished. The base and motor housing are stamped steel with a three-speed slide switch. This fan will oscillate and was a top-of-the-line fan when produced. This may have been a take off on residential fans of the period.

Grade I – $150.00
Grade II – $113.00
Grade III – $75.00
Grade IV – $53.00
Grade V – $30.00
Grade VI – $8.00

Left: ***R & M***
Type: 5104
Blades: 10"
Date: c.1934

A small one-speed non-oscillating fan having a cast-iron base and a steel S style guard. The badge is brass and black with the R&M flag.

Grade I –	$40.00
Grade II –	$23.00
Grade III –	$20.00
Grade IV –	$14.00
Grade V –	$8.00
Grade VI –	$2.00

Center: ***R & M***
Type: 2840
Blades: 9"
Date: c.1918

A small two-speed fan with polished brass windmill Type blades. The universal motor is non-oscillating.

Grade I –	$50.00
Grade II –	$38.00
Grade III –	$25.00
Grade IV –	$18.00
Grade V –	$10.00
Grade VI –	$3.00

Right: ***R & M***
Type: 3100
Blades: 10"
Date: c.1919

An early small oscillator with a universal motor. The base and motor housing are stamped steel. There is a three-speed slide switch in the base. What makes this fan unusual is the small size, 12" and 16 " examples are very common.

Grade I –	$80.00
Grade II –	$60.00
Grade III –	$40.00
Grade IV –	$28.00
Grade V –	$12.00
Grade VI –	$4.00

Robbins & Meyers made some of the most outstanding Art Deco fans during the thirties and early forties. One of their first experiments was the radio fan that got its name because it resembled a radio speaker. In 1937, they came out with their first overlapping blade fan for commercial use. This fan was very similar to an Emerson that came out about the same time. By 1940, R & M had completely streamlined their desk fans in a highly Art Deco style.

Left: ***R & M***
Type: 5404
Blades: 10"
Date: c.1939

An early Art Deco fan with overlapping blades. The base is streamlined and finished in gloss black with sliver trim. A one-speed toggle switch is located at the back of the base. The motor is finished in gloss black with an oscillator mechanism located off the rear bearing. The guard is formed steel wire in a double diamond style with a hexagonal flag badge in black and silver. The fan is very quiet to operate and efficient.

Grade I – $90.00
Grade II – $68.00
Grade III – $45.00
Grade IV – $32.00
Grade V – $18.00
Grade VI – $5.00

Center: ***R & M***
Type: 6000 Radio Fan
Blades: 10"
Date: c.1936

One of the most Art Deco stylized fans ever manufactured. This fan is somewhat similar to the Silver Swan. The motor and blades are supported by the guard giving a totally new effect. The base is made of cast iron and the guard out of stamped steel. The fan is finished in a dark brown, almost a wood tone. The example pictured is a non-oscillator. They also came with an oscillator. For all of the style this fan has, it was rather inefficient and noisy compared to the Silver Swan or GE quiet fan.

Grade I – $150.00
Grade II – $113.00
Grade III – $75.00
Grade IV – $53.00
Grade V – $30.00
Grade VI – $8.00

Right: ***R & M***
Type: D10-A6-01
Blades: 10"
Date: c.1940

This is a fully developed Art Deco desk fan with deeply pitched overlapping blades, reinforced guard, and redesigned motor and base. Wires have been added at the top and bottom of the guard to give it added strength and protection of the blades. The blades have a large center button covering the hub. The motor housing has been redesigned and is hexagonal. It is trimmed in sliver. The base has been made taller and more streamline. The oscillator has been redesigned and is not very easy to engage.

Grade I – $130.00
Grade II – $96.00
Grade III – $65.00
Grade IV – $46.00
Grade V – $26.00
Grade VI – $7.00

Left: ***R & M***
Type: 1204
Blades: 12"
Date: c.1935

A standard Depression era fan with a modified S wire guard. The base and motor housing are made of stamped steel finished in gloss black. A handle is located on top of the motor for easy transport. The base has a three-speed slide switch.

Grade I – $70.00
Grade II – $53.00
Grade III – $35.00
Grade IV – $25.00
Grade V – $14.00
Grade VI – $3.00

Center: ***R & M***
Type: B16AG-0
Blades: 16"
Date: c.1947

A large post-World War II fan with a modern style guard and rounded blades. The base is cast aluminum with a two-speed rotary switch. The motor is made of cast aluminum in the bullet back style. This fan is finished in a light brown finish.

Grade I – $70.00
Grade II – $53.00
Grade III – $35.00
Grade IV – $25.00
Grade V – $14.00
Grade VI – $3.00

Right: ***R & M***
Type: B12AG-10
Blades: 12"
Date: c.1947

A post-World War II desk fan with a one-speed switch and that will oscillate.

Grade I – $50.00
Grade II – $38.00
Grade III – $25.00
Grade IV – $18.00
Grade V – $10.00
Grade VI – $3.00

Peerless
Type: Bi-polar
Blades: 12"
Date: c.1907

A direct current fan with exposed field coils and brushes. The base is made of cast iron with a three-speed slide switch. The guard is made of formed steel wire in an S style.

Grade I – $5,000.00
Grade II – $3,750.00
Grade III – $2,500.00
Grade IV – $1,750.00
Grade V – $1,000.00
Grade VI – $250.00

Peerless
Type: B Spin-Start
Blades: 12"
Date: 1904

At the turn of the century alternating current induction motors had difficulty starting thus posing a real problem for fan designers. This was not a problem with direct current fans that would always start when attached to a power source. Some early companies like Emerson used a start position on the switch to start the fan manually. Once the fan was going, the operator could then choose a speed of choice. If the fan was left in the start position too long the motor would overheat and burn out the starter winding. Peerless solved this problem by developing a fan with a manual spring start. This spring starter turned the rotor that got the blades moving which allowed the fan to go up to full speed. The operator could then select one of two speeds. The castings on this fan are some of the most ornate on any fan manufactured. The large pancake motor sits on a trunnion attached to a fancy ribbed base. The guard is made of heavy formed brass wire in the S style attached to the motor by cast-iron struts. The blades are made of heavy sheet brass in the windmill style.

Grade I – $9,000.00 | Grade IV – $3,150.00
Grade II – $6,750.00 | Grade V – $1,800.00
Grade III – $4,500.00 | Grade VI – $450.00

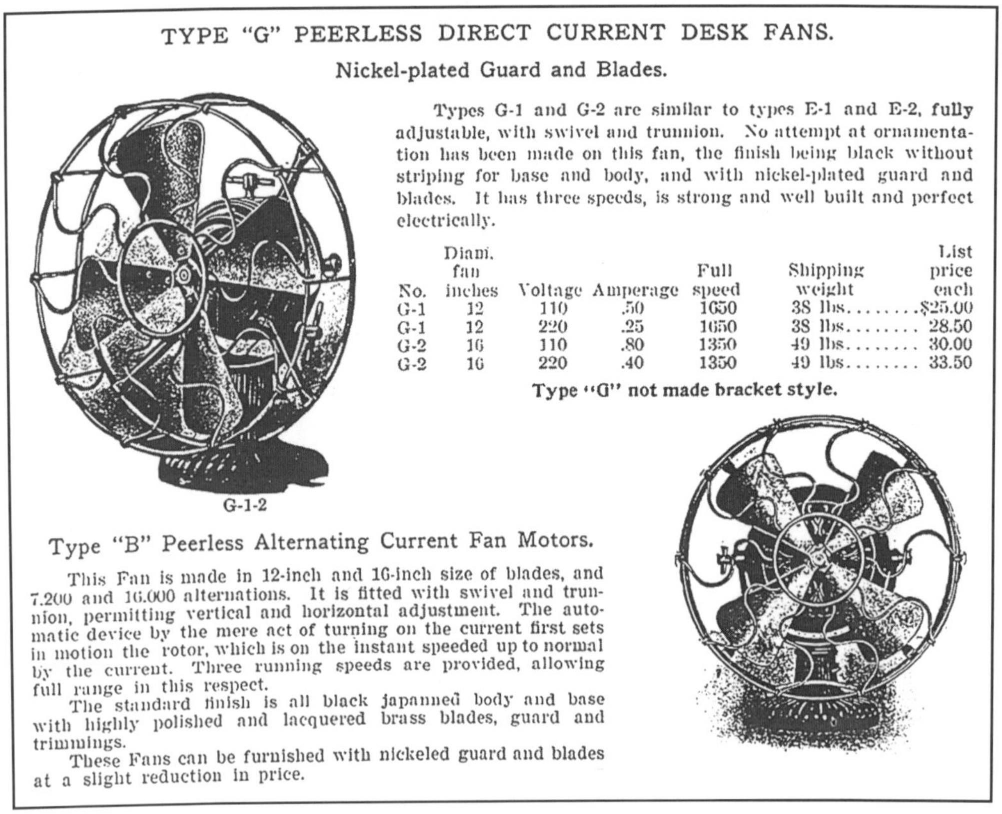

TYPE "G" PEERLESS DIRECT CURRENT DESK FANS.

Nickel-plated Guard and Blades.

Types G-1 and G-2 are similar to types E-1 and E-2, fully adjustable, with swivel and trunnion. No attempt at ornamentation has been made on this fan, the finish being black without striping for base and body, and with nickel-plated guard and blades. It has three speeds, is strong and well built and perfect electrically.

No.	Diam. fan inches	Voltage	Amperage	Full speed	Shipping weight	List price each
G-1	12	110	.50	1650	38 lbs........	$25.00
G-1	12	220	.25	1650	38 lbs........	28.50
G-2	16	110	.80	1350	49 lbs........	30.00
G-2	16	220	.40	1350	49 lbs........	33.50

Type "G" not made bracket style.

G-1-2

Type "B" Peerless Alternating Current Fan Motors.

This Fan is made in 12-inch and 16-inch size of blades, and 7,200 and 16,000 alternations. It is fitted with swivel and trunnion, permitting vertical and horizontal adjustment. The automatic device by the mere act of turning on the current first sets in motion the rotor, which is on the instant speeded up to normal by the current. Three running speeds are provided, allowing full range in this respect.

The standard finish is all black japanned body and base with highly polished and lacquered brass blades, guard and trimmings.

These Fans can be furnished with nickeled guard and blades at a slight reduction in price.

An illustration from a 1904 Illinois Electric Company catalog showing a DC Peerless direct current and a spin-start Peerless fan. These were top-of-the-line products when introduced and very expensive.

Peerless
Type: O DD 1
Blades: 12"
Date: 1914

An example of a direct current front end oscillator. The brush caps protrude from the rear bearing, opposite of Emerson fans. The base and motor housing are cast iron and finished in high gloss black.

Grade I –	$1,700.00
Grade II –	$1,275.00
Grade III –	$850.00
Grade IV –	$596.00
Grade V –	$340.00
Grade VI –	$85.00

Left: ***Peerless***
Type: G-1
Blades: 12"
Date: 1905

Right: ***Peerless***
Type: unknown
Blades: 12"
Date: 1914

An early direct current fan with a heavy cast-iron ribbed base. The guard and blades were originally nickel plated. The motor is held to the base by a large cast-iron trunnion.

Grade I – $4,000.00
Grade II – $2,300.00
Grade III – $2,000.00
Grade IV – $1,400.00
Grade V – $800.00
Grade VI – $200.00

Peerless fans were primarily known for their front oscillators which became a hallmark of their fan design. It seems that having the oscillator away from the spinning blades made sense, since you could cut your finger off. This did not seem to deter Peerless in their fan design. This unusual configuration was soon dropped in the early twenties for a more conventional rear oscillator. The first of these fans had a tab-footed base; the one pictured is a later example. Another feature of Peerless fans is the front guard badge containing all the important information about the fan. Unfortunately, these badges are often missing leaving the collector cluless to the make, model, and type of the fan. The motor has a very modern look and runs quietly.

Grade I – $1,000.00
Grade II – $750.00
Grade III – $500.00
Grade IV – $350.00
Grade V – $200.00
Grade VI – $50.00

Left: ***Peerless***
Type: 240520
Blades: 10"
Date: c.1916

Right: ***R & M***
Type: 42086
Blades: 16"
Date: c.1919

This is a small fan for home use with a front oscillator. The blades are made of polished sheet brass. The base and motor housing are made of cast iron finished in gloss black. The fan has a three-speed slide switch and a universal motor. This would have been comparable to the 10" Northwind made by Emerson.

Grade I – $90.00
Grade II – $68.00
Grade III – $45.00
Grade IV – $32.00
Grade V – $18.00
Grade VI – $5.00

This is an example of a residential six-blade fan. The blades are made of polished brass and the guard is made of formed steel wire in the S style. This fan has a rear oscillator and three speeds.

Grade I – $400.00
Grade II – $230.00
Grade III – $200.00
Grade IV – $140.00
Grade V – $80.00
Grade VI – $20.00

Hunter-Century Gilbert

Century Electric of St. Louis made their first desk fan in 1911. Century placed most of their energy into industrial motors; fans were just a sideline. Their first fans had large open air holes in the front and rear of the motor giving them the name skeletal motors. This design may have been due to industrial motors of the period having large open vent holes for cooling purposes. Fan motors were not under such extreme conditions as factory motors and did not need such large cooling vent openings. In the late thirties Century was purchased by Hunter, hence the name Hunter-Century on later fans.

Left: ***Century***
Type: S-3
Blades: 12"
Date: c.1918

A skeletal non-oscillator having a knuckle-neck between the base and motor. This knuckle-neck was essentially a small trunnion and is a major characteristic of Century skeletal fans. The base and motor housing are made of cast iron. The base contains a five-speed switch that is unusual for fans made after 1908. The blades and guard are made of polished brass.

Grade I – $400.00
Grade II – $230.00
Grade III – $200.00
Grade IV – $140.00
Grade V – $80.00
Grade VI – $20.00

Right: ***Century***
Type: S-3
Blades: 12"
Date: c.1918

A skeletal oscillator with five speeds. The oscillator is very compact and can be engaged by turning a small brass lever on the top of the gear box. These skeletal fans are very quiet when operating.

Grade I – $300.00
Grade II – $225.00
Grade III – $150.00
Grade IV – $105.00
Grade V – $60.00
Grade VI – $15.00

Left: ***Century***
Type: unknown
Blades: 16"
Date: c.1918

A 16" skeletal oscillator with five speeds. The blades are made of sheet brass and have a shallow pitch. The guard is made of formed brass wire in the S style. These fans were primarily made for commercial or industrial use and are very large.

Grade I – $300.00
Grade II – $225.00
Grade III – $150.00
Grade IV – $105.00
Grade V – $60.00
Grade VI – $15.00

Right: ***Century***
Type: Model 14, Type S-3
Blades: 12"
Date: c.1914

An example of a residential six-blade skeletal oscillator. Six-blade Century fans are very unusual and rare. The blades are made of thick sheet brass and tend to resemble GE blades of the same period. The guard is made of formed brass wire in the S style. This fan has five speeds and is very quiet when in operation.

Grade I – $900.00
Grade II – $680.00
Grade III – $450.00
Grade IV – $320.00
Grade V – $180.00
Grade VI – $50.00

Around 1920, Century redesigned their fans with several major changes. The most important feature being the motor was placed in an enclosed case. The first fans from Century had oversized vent holes and their later fans had none. The motor housing on these new fans were made of die-cast metal and did not hold up well. The design looked sleek and modern, but this was misleading. If the fan fell over or was dropped, it was most often ruined. The back plate on the oscillator was especially prone to damage. The later 12" and 16" fans had handles on top of the motor, but they tended to become loose too. The motor badge is almost always missing; falling off due to the die casting.

Left: ***Century***
Type: Model 51
Blades: 9"
Date: c.1920

An early redesigned Century with the totally enclosed motor. The brass guard and blade date this fan as an early example. This fan has three speeds and will oscillate. The examples with brass guards are more collectible.

Grade I – $70.00
Grade II – $53.00
Grade III – $35.00
Grade IV – $25.00
Grade V – $14.00
Grade VI – $3.00

Right: ***Century***
Type: Model 102
Blades: 12"
Date: c.1923

An example of a redesigned Century with a steel guard and brass blades. A handle is located on top of the motor for easy transport. Century continued to use brass blades after most manufacturers discontinued the practice.

Grade I – $80.00
Grade II – $60.00
Grade III – $40.00
Grade IV – $28.00
Grade V – $12.00
Grade VI – $4.00

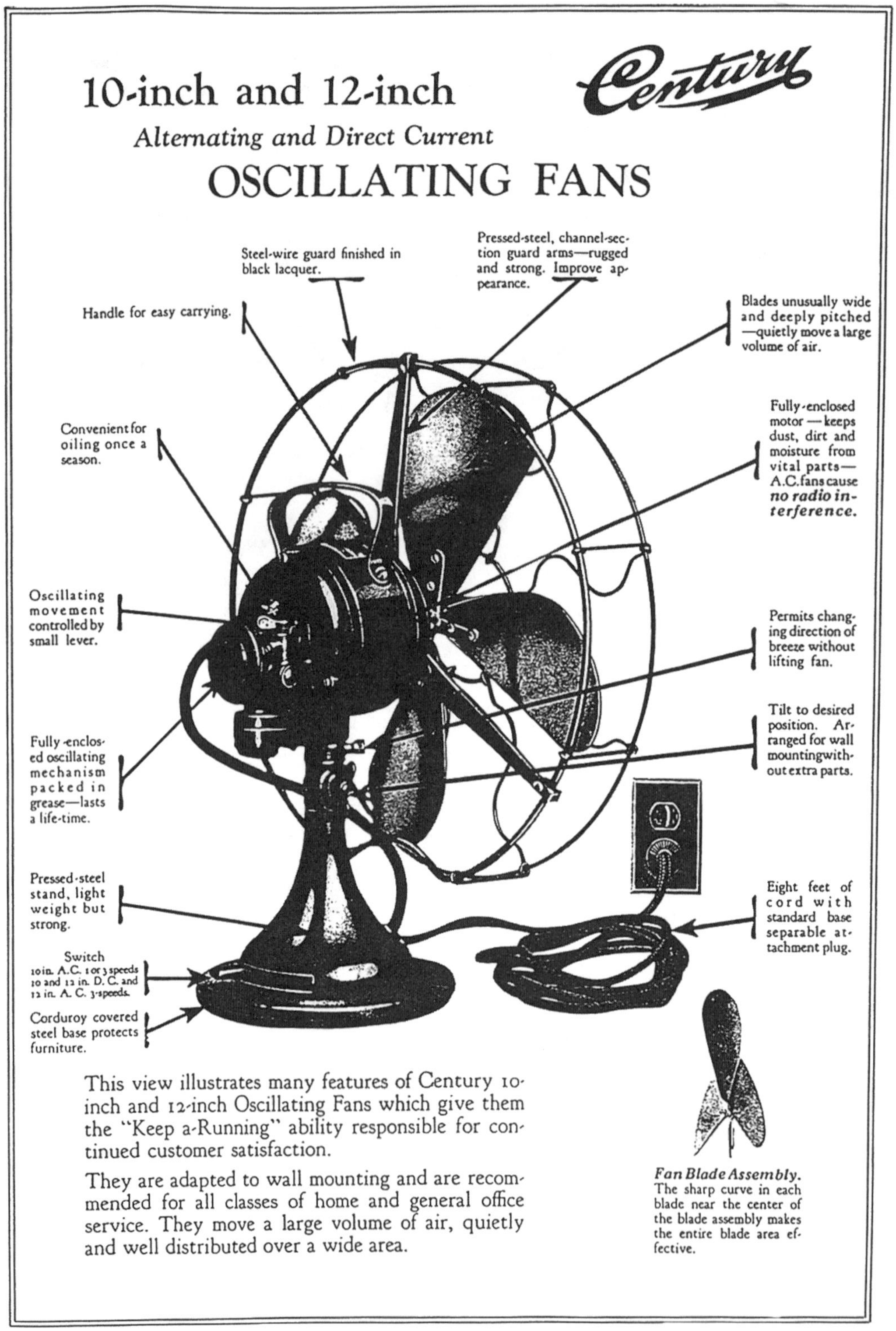

An illustration from the early thirties touting the benefits of Century desk fans. These fans were very sleek and modern in design when introduced in the early twenties. Unfortunately the motor housing and oscillating mechanism were made of die-cast metal that did not hold up well. Century used brass blades on their fans many years after most manufacturers had discontinued the practice. These were not very durable fans.

In 1937 Century redesigned their fans once again to compete with the other fan manufacturers. This new line of fans was called "Zephaire." Unfortunately, some of the things that needed to be changed on their enclosed motor housing were not improved. The motor was essentially left as it had been in the twenties making this new design mostly cosmetic. The blades and guard were completely modified to meet the quiet blade rage of the late thirties. The new cast-iron base was streamlined and contained a two-speed slide switch. The switch is located at the rear of the motor. This new design had serious problems on the 16" fans which tended to have a flimsy guard and blades that were easily knocked out of balance.

Left: ***Century***
Type: Model 601
Blades: 10"
Date: 1937

A newly designed fan with quiet blades and streamlined base. This fan will oscillate and has a handle for easy transport.

Grade I – $50.00
Grade II – $38.00
Grade III – $25.00
Grade IV – $18.00
Grade V – $10.00
Grade VI – $3.00

Right: ***Century/Hunter***
Type: Model 601
Blades: 10"
Date c.1939

This fan is exactly the same as the Century next to it except the badge had changed. Soon after this fan was introduced Hunter, purchased Century and thus the name change.

Grade I – $50.00
Grade II – $38.00
Grade III – $25.00
Grade IV – $18.00
Grade V – $10.00
Grade VI – $3.00

Hunter/Century
Type: Cat #250
Blades: 12"
Date: c.1939

Century also made a line of industrial fans that did not incorporate quiet blades or streamlined bases. These fans had the traditional Century blades and base. However the motor and guard were different and somewhat resembled General Electric. These fans had stamped steel motor housing rather than die cast, making them much superior to the Zephaire fans. This fan has three speeds and will oscillate.

Grade I – $40.00
Grade II – $23.00
Grade III – $20.00
Grade IV – $14.00
Grade V – $8.00
Grade VI – $2.00

Following World War II Hunter/Century manufactured a line of fans for commercial and industrial use. These fans had overlapping polished chrome blades with a bullet back motor and a streamlined base. These fans were produced in large quantities and were good sellers for the company.

Left ***Hunter/Century***
Type: unknown
Blades: 12"
Date: c.1948

An example of a post-World War II Hunter/Century fan. This one is a bit newer than the one on the right. It is chrome plated, has three speeds, and oscillators.

Grade I – $100.00
Grade II – $75.00
Grade III – $50.00
Grade IV – $35.00
Grade V – $20.00
Grade VI – $5.00

Center: ***Gilbert***
Type: unknown
Blades: 12"
Date: c.1936

The A.C. Gilbert company primarily manufactured toys, erector sets, and toy trains, and fans were a sideline. Gilbert fans tended to be very stylish, but not very efficient. This fan has a chrome-plated guard, blades, and motor housing, the base is cast iron with a black wrinkle finish. The guard and badge are highly stylized Art Deco.

Grade I – $150.00
Grade II – $113.00
Grade III – $75.00
Grade IV – $53.00
Grade V – $30.00
Grade VI – $8.00

Right: ***Hunter/Century***
Type: C-12
Blades: 12"
Date: c.1946

An example of a post-World War II Hunter/Century fan. It is chrome plated, has three speeds, and oscillators.

Grade I – $100.00
Grade II – $75.00
Grade III – $50.00
Grade IV – $35.00
Grade V – $20.00
Grade VI – $5.00

Left: ***Gilbert***
Type: unknown
Blades: 8"
Date c.1935

A small fan with highly polished steel blades. The fan is finished in a light green wrinkle finish and will oscillate.

Grade I – $30.00
Grade II – $23.00
Grade III – $15.00
Grade IV – $10.00
Grade V – $6.00
Grade VI – $2.00

Right: ***Gilbert***
Type: unknown
Blades: 10"
Date: c. 1934

A better grade Gilbert fan with three speeds and an oscillator. This example has highly polished chrome blades.

Grade I – $40.00
Grade II – $23.00
Grade III – $20.00
Grade IV – $14.00
Grade V – $8.00
Grade VI – $2.00

Left: ***Gilbert***
Type: M1384
Blades: 12"
Date: c.1935

A very stylish desk fan with a light green wrinkle finish and highly polished chrome blades. The fan has a rather unusual step base with a three-speed slide switch. The motor is quite small and the oscillator is very crudely made.

Grade I –	$70.00
Grade II –	$53.00
Grade III –	$35.00
Grade IV –	$25.00
Grade V –	$14.00
Grade VI –	$3.00

Center: ***Gilbert***
Type: unknown
Blades: 10"
Date: c.1935

A small oscillating desk fan with large chrome blades.

Grade I –	$30.00
Grade II –	$23.00
Grade III –	$15.00
Grade IV –	$10.00
Grade V –	$6.00
Grade VI –	$2.00

Right: ***Gilbert***
Type: unknown
Blades: 12"
Date: c.1937

An Emerson Silver Swan look-a-like. This fan has aluminum overlapping blades surrounded by a chrome-plated guard. The base is cast iron and attached to the motor by a shaft. The blades are made out of very thin aluminum and are quite fragile. The guard seems over built for the fan and is a very unusual Art Deco style.

Grade I –	$240.00
Grade II –	$180.00
Grade III –	$120.00
Grade IV –	$84.00
Grade V –	$48.00
Grade VI –	$12.00

Left: ***Gilbert***
Type: A-325
Blades: 10"
Date: c.1946

Following World War II Gilbert manufactured this style fan in 8", 10", and 12" models. The base is made of cast aluminum and the motor housing is made of stamped steel. The blades and guard are made of polished chrome with a large central button over the center of the blade. This fan has a small rotary switch on the back of the bullet and will oscillate. The fan is finished in light blue.

Grade I – $30.00
Grade II – $23.00
Grade III – $15.00
Grade IV – $10.00
Grade V – $6.00
Grade VI – $2.00

Right: ***Gilbert***
Type: A-54
Blades: 12"
Date: 1940

In 1940, Gilbert introduced their new streamlined Art Deco fan. Like the General Electric quiet fan, this model was very stylish and attractive. All aspects of the fan had been redesigned to look modern and up to date. The base is made of cast iron and the motor housing is stamped steel in an unusual elliptical shape. The blades are large and overlapping. The guard is made of formed steel wire painted white. The fan has two speeds and will oscillate.

Grade I – $270.00
Grade II – $203.00
Grade III – $135.00
Grade IV – $95.00
Grade V – $54.00
Grade VI – $14.00

Wagner/Galvin/Knapp-Monarch

Wagner, Galvin, and Knapp-Monarch were all electric companies headquartered in St. Louis, Missouri. Wager was the largest of the three with a large factory located at Wellston, Missouri, just outside of the St. Louis city limits. Wagner manufactured many electrical related devices as well as motors. Fans were a sideline and not a major manufacturing concern. Wagner started making fans in the early 1890s and are very collectible fans. When the Depression came, Emerson was almost wiped out due to the drop in fan sales, but this was not the case with Wagner being more diversified. Knapp-Monarch was another electrical manufacturer located in St. Louis and began the production of electrical motors at the turn of the century. It seems that this company made motors for several other small fan companies and is best known for their deco style fans. Galvin Electric was located near the Anheuser-Busch brewery in South St. Louis. They seemed to have used other manufacturer's parts to make their fans. They made fans around World War I and then seemed to have quit manufacturing fans altogether.

Left: ***Wagner Electric***
Type: 9260
Blades: 10"
Date: c.1919

A small fan made for residential home use. The blades are made of polished sheet brass and the guard is formed steel wire in the S style. The fan is finished a gloss back with a one-speed slide switch in the base. This fan will oscillate.

Grade I – $60.00
Grade II – $45.00
Grade III – $30.00
Grade IV – $21.00
Grade V – $12.00
Grade VI – $3.00

Center: ***Wagner Electric***
Type: 6260
Blades: 12"
Date: c.1930

This fan is very well made and is attractive. The blades and guard are made of steel. The motor housing and base are made of cast iron. On top of the motor housing is a handle for easy transport. The base contains a three-speed slide switch and the fan will oscillate.

Grade I – $60.00
Grade II – $45.00
Grade III – $30.00
Grade IV – $21.00
Grade V – $12.00
Grade VI – $3.00

Right: ***Wagner Electric***
Type: 62601
Blades: 12"
Date: c.1935

This fan is very similar to Type 6260 with a few changes. The base is newly designed in a streamlined Art Deco style. The large slide switch has been replaced by a compact two-speed switch. The motor is spot welded to a bracket that attaches to the base. The guard, blades, and motor have not been changed from fans made by Wagner in the early thirties. This has an excellently well designed motor that was used on all of their fans of the period.

Grade I – $70.00
Grade II – $53.00
Grade III – $35.00
Grade IV – $25.00
Grade V – $14.00
Grade VI – $3.00

Left: ***Wagner Electric***
Type: 52603
Blades: 10"
Date: c.1939

Wagner redesigned their line of desk fans in the late thirties to meet the demand for streamlined products. This fan has newly designed overlapping quiet. Type blades with a deco steel guard. The Art Deco base is made of cast iron and has silver trim. A small compact three-speed slide switch is located on the right side of the base. This fan is very quiet and will oscillate.

Grade I – $70.00
Grade II – $53.00
Grade III – $35.00
Grade IV – $25.00
Grade V – $14.00
Grade VI – $3.00

Right: ***Wagner Electric***
Type: 5260
Blades: 10"
Date: c.1931

One of the last fans made by Wagner before they began to redesign the base. The base and blades are made of stamped steel and the motor housing is cast iron. The fan will oscillate and has a three-speed switch in the base.

Grade I – $50.00
Grade II – $38.00
Grade III – $25.00
Grade IV – $18.00
Grade V – $10.00
Grade VI – $3.00

In 1940 Knapp-Monarch came out with a newly designed line of fans for consumer use. These fans had a sleek modern style base, guard, and blades. The motor was covered by a bullet back hiding the oscillating mechanism. The fan was finished in a dark brown wrinkle finish with chrome highlights. This fan became a prototype of fans made by Knapp-Monarch all through the forties. The pre-war examples have a cast-iron base, the post-war examples have cast aluminum.

Left: ***Knapp-Monarch***
Type: unknown
Blades: 9"
Date: 1940

A small non-oscillating fan with a brown wrinkle finish.

Grade I – $30.00
Grade II – $23.00
Grade III – $15.00
Grade IV – $10.00
Grade V – $6.00
Grade VI – $2.00

Right: ***Knapp-Monarch***
Type: Model 3-503
Blades: 12"
Date: 1940

A completely redesigned desk fan with overlapping blades and a modern streamlined base, blades, and guard. This fan has a brown wrinkle finish.

Grade I – $70.00
Grade II – $53.00
Grade III – $35.00
Grade IV – $25.00
Grade V – $14.00
Grade VI – $3.00

Left: ***Galvin***
Type: A-F-10
Blades: 10"
Date: c.1920

This fan incorporates many parts manufactured by Emerson. The base and motor are very characteristic of Emerson Junior fans of the early twenties. The base has a toggle switch instead of a slide switch but for all other intents and purposes is an Emerson Junior base. The motor has large daisy style air vents that are more characteristic of Trojan fans of the period. The single bearing configuration gives it away that it is a true Emerson motor. There is a large cast handle on top of the motor for easy transport. The guard is exactly like a 10" Emerson Junior guard but with eight wires instead of six. The guard badge is an Emerson badge with the Galvin name. A very unusual and attractive fan.

Grade I – $110.00
Grade II – $83.00
Grade III – $55.00
Grade IV – $39.00
Grade V – $22.00
Grade VI – $6.00

Right: ***Galvin***
Type: UGF-10
Blades: 10"
Date: c.1919

This fan has a universal motor and may have been made by Emerson. The base is a standard Century Electric base that does not fit very well with the motor. The guard is an Emerson product. This fan does not fit together very well and looks rather awkward.

Grade I – $110.00
Grade II – $83.00
Grade III – $55.00
Grade IV – $39.00
Grade V – $22.00
Grade VI – $6.00

StarRite/Eskimo

StarRite fans were manufactured by the Fitzgearld Company which made low priced dime store fans. Some general characteristics that make these fans interesting are that they used stamped steel motor housings and bases early in their manufacturing like Westinghouse. Some of their fans are all nickel or chrome plated which is an unusual finish for a small dime store fan. Their construction is not very durable and many of these fans did not last. For example, the base on many of these fans were pop riveted together making it impossible to replace the cord without destroying the base. This was the prototype of the throw away fans which are manufactured today.

Left: ***StarRite***
Type: unknown
Blades: 8"
Date: c.1918

A small all-chrome one-speed fan with stamped steel parts. The slits in the base are very characteristic of StarRite fans.

Grade I – $40.00
Grade II – $23.00
Grade III – $20.00
Grade IV – $14.00
Grade V – $8.00
Grade VI – $2.00

Center: ***StarRite***
Type: unknown
Blades: 10"
Date: c.1918

This design was used by StarRite during the late teens and early twenties. The early examples have a universal motor, but this was changed in the twenties. The windmill type blades give the fan an older look. The fan is finished in polished nickel plate and will oscillate. The guard is rather flimsy and did not attach to the motor in a very secure manner.

Grade I – $60.00
Grade II – $45.00
Grade III – $30.00
Grade IV – $21.00
Grade V – $12.00
Grade VI – $3.00

Right: ***StarRite***
Type: unknown
Blades: 8"
Date: c.1922

A later example with an alternating current only motor and redesigned guard struts. The new guard struts are improved but spot welded to the guard in a poor fashion. A one-speed non-oscillating dime store fan.

Grade I – $30.00
Grade II – $23.00
Grade III – $15.00
Grade IV – $10.00
Grade V – $6.00
Grade VI – $2.00

Left:
Type: Fitzgerald
Blades: 10"
Date: c.1947

This is a very unusual post-World War II fan with modern characteristics. The motor is held to the base by a chrome ring that can be adjusted up or down. The base and motor have been streamlined in a modern fashon. The base contains a three-speed toggle switch. The guard and blades are finished in polished chrome and the base and motor are light brown. The fan is rather noisy and not very efficient, but is unusual in design.

Grade I – $70.00
Grade II – $53.00
Grade III – $35.00
Grade IV – $25.00
Grade V – $14.00
Grade VI – $3.00

Right: ***StarRite***
Type: unknown
Blades: 10"
Date: c.1933

This fan has several Art Deco aspects. The octagonal guard is unusual and rather striking. The blades are polished brass and contrast with the lime green gloss finish of the base and motor housing. The base is octagonal with a three-speed slide switch. This fan will oscillate.

Grade I – $110.00
Grade II – $83.00
Grade III – $55.00
Grade IV – $39.00
Grade V – $22.00
Grade VI – $6.00

Berstead Manufacturing

Berstead Manufacturing of Ohio made many low priced quality fans after World War II. They went under such trade names as Eskimo, Zero, and Wizard. These fans were mass produced in large quantities to meet the pent-up demand for consumer goods after World War II. These fans were very stylish and attractive to consumers. Unfortunately, they had one major drawback: a poorly designed oscillator. The oscillator was located at the back of the fan hidden by a bullet cover. To engage the oscillator you had to tighten a screw clamp that would use friction on the oscillator shaft to cause rotation. This looked easy to operate, but didn't work as well as planned. More often than not, the screw would not hold for a long period of time and the fan would stop oscillating. If you didn't want the fan to oscillate there was no good way to keep the fan stationary without having it tilted to one side. This made the fan look awkward. The bearings were very poor and if the fan was not oiled regularly it would screech loudly. One redeeming feature of these fans is that they were inexpensive to purchase at a time when fans were in great demand.

Left: ***Berstead Manufacturing***
Type: Wizard, Model 45-M
Blades: 12"
Date: c.1946

This fan has post-war modern design features such as a cast-iron streamlined ribbed base, bullet back motor, and a cross guard. The guard is especially attractive on this fan. The blades are made of polished aluminum with a small center button. The motor housing and base are finished in a black wrinkle paint, with a highly polished chrome bullet at the rear of the motor. This fan has three speeds.

Grade I – $50.00
Grade II – $38.00
Grade III – $25.00
Grade IV – $18.00
Grade V – $10.00
Grade VI – $3.00

Center: ***Berstead Manufacturing***
Type: Eskimo, Model 1005
Blades: 10"
Date: c.1947

This fan is a non-oscillator model with an unusual guard design. The guard forms a star and is made of polished chrome-plated formed steel wire. The front section of the motor is finished in a black wrinkle paint, the rear is covered by a large chrome bullet. The base is cast iron with a chrome Eskimo badge in the front.

Grade I – $30.00
Grade II – $23.00
Grade III – $15.00
Grade IV – $10.00
Grade V – $6.00
Grade VI – $2.00

Right: ***Berstead Manufacturing***
Type: Eskimo, Model 45-K
Blades: 12"
Date: c.1946

This is the most common fan made by Bersted Manufacturing after the war. The guard is made of chrome-plated formed steel wire with an outer ring in the shape of a cobweb, hence the name cobweb guard. The blades are polished aluminum. The front of the motor is finished in black wrinkle paint, the rear covered with a large polished chrome bullet. The base is cast iron with a three-speed rotary switch and this example will oscillate.

Grade I – $40.00
Grade II – $23.00
Grade III – $20.00
Grade IV – $14.00
Grade V – $8.00
Grade VI – $2.00

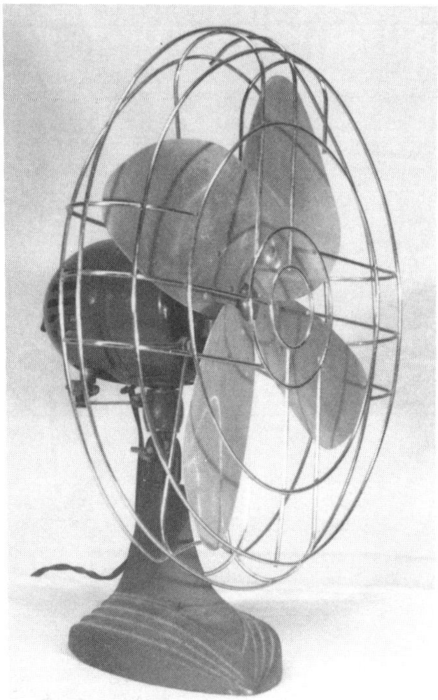

Berstead Manufacturing
Type: Wizard
Blades: 16"
Date: c.1949

A large post-World War II fan with stylish features. The base is made of cast iron with streamlined ribbing. There are no electrical parts in the base. The motor housing is made of stamped steel with a large bullet covering the oscillation mechanism. A two-speed rotary switch is located at the rear of the bullet. The guard is made of thick formed steel wire in a modern cross style. The blades are made of sheet aluminum and attached to a steel spider. The blades rotate counter-clockwise like Westinghouse fans. The motor is finished in a metallic blue color and is very sleek in design.

Grade I – $70.00
Grade II – $53.00
Grade III – $35.00
Grade IV – $25.00
Grade V – $14.00
Grade VI – $3.00

Berstead Manufacturing
Type: Eskimo, Model 55-D
Blades: 16"
Date: c.1935

Bersted also made fans before World War II and this is an example. This is a large commercial grade fan with an unusual step base. The motor and base are made of stamped steel and finished in gloss black with silver trim. The guard has a stylized S configuration finished in polished chrome. The blades are similar to early Westinghouse fans and they are finished in polished chrome. This fan will oscillate.

Grade I – $70.00
Grade II – $53.00
Grade III – $35.00
Grade IV – $25.00
Grade V – $14.00
Grade VI – $3.00

Right: **Berstead Manufacturing**
Type: Eskimo
Blades: 16"
Date: c.1946

A large commercial cobweb guard fan. The larger fans have a very difficult time with the oscillator staying engaged and this was a weak point on this fan. The blades and guard are finished in polished chrome. The base and motor are finished in black wrinkle paint. A large chrome bullet covers the rear of the motor.

Grade I – $70.00
Grade II – $53.00
Grade III – $35.00
Grade IV – $25.00
Gradc V – $14.00
Grade VI – $3.00

Berstead Manufacturing
Type: Eskimo
Blades: 18"
Date: c.1946

This is the king of all cobweb guard fans. This commercial fan was designed for use in a store or other large building. The blades are made of a single peace of cast aluminum with a large highly polished chrome bullet. The motor seems to float in mid air, but is actually attached to a strap steel trunnion attached to the spaceship-looking base. The base is cast iron finished in a wrinkle black paint with chrome trim. This fan was very futuristic when manufactured.

Grade I – $130.00
Grade II – $98.00
Grade III – $65.00
Grade IV – $46.00
Grade V – $26.00
Grade VI – $7.00

Western Electric

Western Electric did not manufacture electric fans; they just marketed them. They were in the communication business and manufactured telephones and telegraph equipment on a large scale. As a large wholesaler of electric equipment, they began selling fans manufactured by other companies with their name on it. This can be very confusing to a collector when they find a Westinghouse, GE, or Robbins & Meyers fan with a Western Electric tag.

Left: ***Western Electric***
Type: W-134058-A (made by Westinghouse)
Blades: 12"
Date: 1912

This is a Westinghouse vane oscillator with a Western Electric Hawthorn guard and motor badge. It is exactly the same as a Westinghouse in every regard except for the badge. The motor and base are made of cast iron. The blades, guard, and vane are made of brass. The large center Hawthorn badge is made of cast brass and is very impressive.

Grade I –	$1,100.00	Grade IV –	$385.00
Grade II –	$825.00	Grade V –	$220.00
Grade III –	$550.00	Grade VI –	$55.00

Right: ***Western Electric***
Type: 105452A (made by Robbins & Meyers)
Blades: 12"
Date: 1909

This fan is commonly referred to as a lollipop oscillator due to the shape of the vane. These fans were manufactured by Robbins & Meyers and are an example of an early vane oscillator. The large tank motor and tab base are made of cast iron and finished in a gloss black. The blades and guard are polished brass, as well as the vane. This very unusual configuration worked quite well, but was abandoned after Robins & Meyers perfected their mechanical oscillator.

Grade I –	$4,000.00	Grade IV –	$1,400.00
Grade II –	$3,000.00	Grade V –	$800.00
Grade III –	$2,000.00	Grade VI –	$200.00

Early direct current fans occasionally had open exposed field coils in a C shape and are commonly called bi-polar fans. This is rather misleading because all electric motors have two poles in order to operate. These fans look more like small generators than fans. Western Electric marketed several of these fans, but did not manufacture them.

Right: ***Western Electric***
Blades: 12"
Date: c.1908

A direct current fan with exposed field coils and brushes. The motor is attached to a cast-iron tab-footed base with a three-speed switch. The blades are made of sheet brass and the guard is made of formed brass wire. The center badge is cast brass with the name "Western Electric."

Grade I –	$5,000.00	Grade IV –	$1,750.00
Grade II –	$3,750.00	Grade V –	$1,000.00
Grade III –	$2,500.00	Grade VI –	$250.00

Left: ***Western Electric***
Type: D06206 (perhaps made by Diehl)
Blades: 12"
Date: c.1917

A residential oscillator with six highly polished brass blades. The guard is also brass with a black and brass badge. The motor and base are finished in gloss black. The fan has three speeds and will oscillate.

Grade I – $700.00
Grade II – $525.00
Grade III – $350.00
Grade IV – $245.00
Grade V – $140.00
Grade VI – $35.00

Right: ***Western Electric***
Type: List #6344 (made by Robbins & Meyers)
Blades: 12"
Date: c.1919

A six-blade residential oscillator with a stamped steel motor housing and base. This is a standard Robbins & Meyers fan with a Western Electric badge. Robbins and Meyers made many fans for Western Electric in the late teens and early twenties. Six-blade examples are rare, while four-blade are common.

Grade I – $600.00
Grade II – $450.00
Grade III – $300.00
Grade IV – $210.00
Grade V – $120.00
Grade VI – $30.00

Polar Cub

Polar Cub fans were made for the dime store trade. The early Polar Cub fans were very small, but later, 10" fans became common. Like the StarRite fans, Polar Cub fans were quite inexpensive when new and not made for commercial or industrial use.

Left: ***Polar Cub***
Type: E
Blades: 5"
Date: c.1918

One of the very first fans made by Polar Club. The guard with the large open ring makes this fan very unusual. The button at the center of the blades is stamped with "Polar Cub" and is made of aluminum. The universal motor is attached to a cast-iron base.

Grade I – $130.00
Grade II – $89.00
Grade III – $65.00
Grade IV – $46.00
Grade V – $26.00
Grade VI – $7.00

Center: ***Polar Cub***
Type: D
Blades: 5"
Date: c.1919

This fan has a nickel-plated universal motor with an S style cage. The base is cast iron and there is a one-speed slide switch.

Grade I – $100.00
Grade II – $75.00
Grade III – $50.00
Grade IV – $35.00
Grade V – $20.00
Grade VI – $5.00

Right: ***Polar Cub***
Type: G
Blades: 5"
Date: c.1921

This fan has a ribbed cast-iron base and the blades are painted gold. A one-speed plug-in fan.

Grade I – $100.00
Grade II – $75.00
Grade III – $50.00
Grade IV – $35.00
Grade V – $20.00
Grade VI – $5.00

Left: ***Polar Cub***
Type: H
Blades: 8"
Date: c.1922

A small non-oscillating fan with a universal motor. The base is cast iron with an open S wire guard.

Grade I – $50.00
Grade II – $38.00
Grade III – $25.00
Grade IV – $18.00
Grade V – $10.00
Grade VI – $3.00

Center: ***Polar Cub***
Type: D
Blades: 9"
Date: c.1919

A small non-oscillating fan with a nickel-plated universal motor. The blades are made of stamped steel painted silver. The base is cast iron and has a two-speed slide switch.

Grade I – $60.00
Grade II – $45.00
Grade III – $30.00
Grade IV – $21.00
Grade V – $12.00
Grade VI – $3.00

Right: ***Polar Cub***
Type: unknown
Blades: 10"
Date: c.1925

During the middle and late twenties Polar Cub manufactured large quantities of dime store grade fans. These fans were made of stamped steel and cheaply constructed. This example has a one-speed slide switch in the base that is very poorly made. There is an oscillator on the back of the fan. Polar Cub placed a decal on top of the motor housing with the type, volts, and other information. Unfortunately these decals became soiled and did not survive the test of time.

Grade I – $30.00
Grade II – $23.00
Grade III – $15.00
Grade IV – $10.00
Grade V – $6.00
Grade VI – $2.00

Left: ***Polar Cub***
Type: unknown
Blades: 10"
Date: c.1932

During the early thirties Polar Cub began to streamline their guard and base. This example has a cast-iron streamlined ribbed base with a modified S guard. The blades and motor house are made of stamped steel. The fan is finished in a light tan with polished chrome blades.

Grade I – $40.00
Grade II – $23.00
Grade III – $20.00
Grade IV – $14.00
Grade V – $8.00
Grade VI – $2.00

Center: ***Polar Cub***
Type: unknown
Blades: 10"
Date: c.1925

A non-oscillating inexpensive dime store fan. The blades, motor housing and base are all made of stamped steel. The motor is very small and poorly constructed.

Grade I – $30.00
Grade II – $23.00
Grade III – $15.00
Grade IV – $10.00
Grade V – $6.00
Grade VI – $2.00

Right: ***Polar Cub***
Type: unknown
Blades: 10"
Date: c.1930

This fan is two tone with a light brown motor and black base. It is an oscillator with a toggle switch in the base.

Grade I – $30.00
Grade II – $23.00
Grade III – $15.00
Grade IV – $10.00
Grade V – $6.00
Grade VI – $2.00

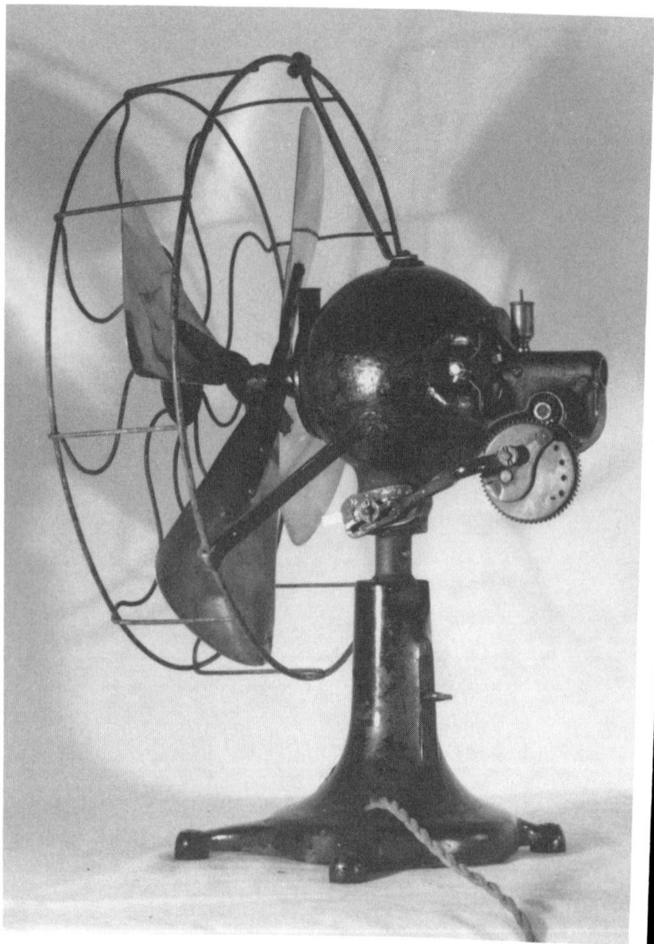

Charles Eck was an early innovator in electric fans. He first patented an oscillator in 1903, and went on to produce high quality desk fans. His most famous fan was the Eck Hurricane. Eck fans were not produced in large quantities like Emerson or General Electric and are quite rare. The fan pictured above is a direct current round ball Eck Hurricane oscillator. The base and motor are made of cast iron. The brushes are at the rear of the of the motor and are exposed. There are two large oil cups located at each bearing. The oscillator is large and exposed. It is a hallmark of Eck oscillating fans. The blades are made of sheet brass and are off center. This blade is referred to as a ming blade and is also a hallmark of Eck fans. The guard is made of formed steel wire in the S style. The badge is polished brass. The base sits on four tab feet and has a slide switch in the front.

Grade I – $5,000.00
Grade II – $3,750.00
Grade III – $2,500.00
Grade IV – $1,750.00
Grade V – $1,000.00
Grade VI – $250.00

Left: ***Air-Castle***
Type: Model 534
Blades: 14"
Date: c.1947

Right: ***Fresh'ND-Aire***
Blades: 8"
Date: 1949

This is a post-World War II fan with modern styling. The base is made of sheet steel having a black wrinkle finish. The bottom section of the base has an outer and inner ring of highly polished chrome giving it a futuristic look. The motor is small with a chrome-plated bullet on the back. The guard is made of heavy formed steel chrome-plated wire and looks very industrial in design. The blade is a one-piece cast-aluminum blade in an airplane configuration and is not very quiet when in operation. This fan has an industrial post-war look.

Grade I – $50.00
Grade II – $38.00
Grade III – $25.00
Grade IV – $18.00
Grade V – $10.00
Grade VI – $3.00

This fan incorporates the latest in post-war design with a completely stylized look. The base is made of dark brown Bakelite plastic and the guard is formed steel which pivots up and down. The blades are deep pitched and pointed. This fan is designed to blend in with the most modern post-war furnisings.

Grade I – $70.00
Grade II – $53.00
Grade III – $35.00
Grade IV – $25.00
Grade V – $14.00
Grade VI – $3.00

Fresh'NDAire
Type: Model 17R
Blades: 24"
Date: c.1950

Fresh'NDAire primarily manufactured large fans for commercial and industrial use. The blades on these fans rotate at a high velocity and have a very characteristic sound. The base, guard, and motor housing are made of steel finished in highly polished chrome. The rear of the motor has been covered by a large bullet giving it a streamlined appearance. The switch is a pull-chain type with a small window showing which speed the fan is operating. This switch is an unusual configuration and is not as efficient as a rotary switch would have been. This is a very lightweight fan for being such a large size.

Grade I – $70.00
Grade II – $53.00
Grade III – $35.00
Grade IV – $25.00
Grade V – $14.00
Grade VI – $3.00

The Dominion fan company made very stylized inexpensive dime store fans in the thirties and forties. These were modestly priced fans that were quite efficient as well as stylish. The ribbing on the blades is a hallmark of Dominion fans.

Left: ***Dominion***
Type: Model 194
Blades: 10"
Date: c.1938

Stylized Art Deco fan with chrome blades and guard, cast-iron ribbed base, and a small non-oscillating motor. The fan is finished in black wrinkle paint with a large button over the blade hub which is very characteristic of Dominion fans.

Grade I – $70.00
Grade II – $53.00
Grade III – $35.00
Grade IV – $25.00
Grade V – $14.00
Grade VI – $3.00

Center: ***Dominion***
Type: Model 932
Blades: 10"
Date: 1940

In 1940, Dominion came out with a line of Art Deco fans with a shield shaped guard. This was an interesting configuration and was done for cosmetic purposes. The base is made of cast iron with ribbing. The motor has a bullet on the back to cover the oscillator. This was a top-of-the-line fan when manufactured. The fan is finished in a black wrinkle paint and the guard and blade are highly polished chrome.

Grade I – $60.00
Grade II – $45.00
Grade III – $30.00
Grade IV – $21.00
Grade V – $12.00
Grade VI – $3.00

Right: ***Dominion***
Type: Model 2017
Blades: 12"
Date: c.1952

In the last years of Dominion fans they dropped cast-iron bases for stamped steel. The large button over the hub and ribbed blades were still incorporated in the design. The bullet at the rear of the fan has been rounded and has a two-speed rotary switch. The fan is finished in a light blue color and is very attractive.

Grade I – $40.00
Grade II – $23.00
Grade III – $20.00
Grade IV – $14.00
Grade V – $8.00
Grade VI – $2.00

Right: ***Dominion***
Type: unknown
Blades: 12"
Date: c.1946

This fan has a cast-iron base finished in a light blue wrinkle color. There is a small red decal across the front of the base with the name of the fan, this is often missing. The motor housing is stamped steel with a two-speed rotary switch. The blades are made of stamped steel with ribbing and finished in a light blue color. The guard is modern in design and painted silver. This fan will oscillate.

Grade I – $50.00
Grade II – $38.00
Grade III – $25.00
Grade IV – $18.00
Grade V – $10.00
Grade VI – $3.00

Left: ***Dominion***
Type: Model 55
Blades: 16"
Date: c.1938

This all chrome fan is very dynamic in appearance. The base and motor housing are made of stamped steel plated in highly polished chrome. The blades are large and overlapping with ribbing. The guard is deco in design and painted silver. This fan has three speeds and will oscillate. This fan was most likely marketed for commercial use.

Grade I – $300.00
Grade II – $225.00
Grade III – $150.00
Grade IV – $105.00
Grade V – $60.00
Grade VI – $15.00

Barber-Coleman manufactured small dime store fans in the middle thirties. These fans were unusual because the base and motor housing were made of Bakelite, an early plastic.

Left: ***Barber-Coleman***
Blades: 10"
Date: c.1935

This was the largest fan made by Barber-Coleman having a bakelite base and motor housing. The fan is very lightweight and has stamped aluminum blades. The guard is made of aluminum wire in the S style with a small Bakelite badge. The motor is very compact and has a rotary switch on the top of the back of the motor. This fan will oscillate.

Grade I – $90.00
Grade II – $68.00
Grade III – $45.00
Grade IV – $32.00
Grade V – $18.00
Grade VI – $5.00

Right: ***Barber-Coleman***
Blades: 8"

Two blade fans are unusual except on dime store fans of the thirties. This fan has small stamped aluminum blades attached to the motor off-center. The base and motor housing are made of brown Bakelite.

Grade I – $70.00
Grade II – $53.00
Grade III – $35.00
Grade IV – $25.00
Grade V – $14.00
Grade VI – $3.00

The Diehl fan company manufactured high quality fans for many years. They also manufactured fans for other companies under the other companies trade name. This practice was established early in the industries development. The picture above illustrates two examples.

Left: ***Singer "Ribbonaire"***
Type: Model G-1
Blades: 6"
Date: 1934

This fan is very unusual because the blades are made of cloth ribbons and the base out of Bakelite. It was marketed by the Singer Sewing Machine Company, but actually manufactured by Diehl. This fan represents a complete break away from the traditional desk fan design and is very innovative. Singer marketed the fan by saying it was much safer than conventional fans and perfect for use around children. They went on to say you could place perfume on the ribbons scenting the air as well as keeping you cool. This was an interesting marketing idea. This seemed to have worked because they were rather popular fans at the time.

Grade I – $150.00
Grade II – $113.00
Grade III – $75.00
Grade IV – $53.00
Grade V – $30.00
Grade VI – $8.00

Center: ***Command-air***
Type: 2405
Blades: 10"
Date: c.1936

Command-air fans are unusual because they have six-blades. By the mid-thirties residential six-blade fans had fallen out of favor with fan manufactures, not even Emerson was making them. This is a very cheaply made fan with stamped steel blades, base, and motor housing. The slide switch is very poor and is made out of a cardboard like substance. The guard is the S type with a brass and green center badge with a large C on it. The information about the fan is found on a cardboard plate underneath the base. The fan is finished in a flat hunter green.

Grade I – $50.00
Grade II – $38.00
Grade III – $25.00
Grade IV – $18.00
Grade V – $10.00
Grade VI – $3.00

Right: ***Diehl***
Type: E-12012
Blades: 12"
Date: c.1918

This fan has a very heavy cast-iron base and motor housing. The spoke guard is very modern looking and commonly found on Diehl fans of the period. This is a very quiet fan when in operation. This example is a non-oscillator, but they did come with oscillators.

Grade I – $60.00
Grade II – $45.00
Grade III – $30.00
Grade IV – $21.00
Grade V – $12.00
Grade VI – $3.00

An illustration of a 1927 ad for a Diehl fan. Diehl made many high quality fans for home and commercial use. The fan pictured was designed around World War I and manufactured until the late twenties. The spoke guard was a hallmark of Diehl fans all through the twenties.

During the thirties much energy went into designing new and different applications for desk fans. New materials made available through technology made this possible. Many of these new composition fans made during the Depression are very collectible as well as interesting.

Left: ***Sampson***
Type: "SafeFlex" No. 710
Blades: 10"
Date: c.1936

To attract consumers to purchase fans, Sampson developed a new prototype fan that had rubber blades. The big selling point of these new fans was safety. The advancement of chemical technology made rubber much cheaper and easier to manufacture making this design possible. The deeply pitched blades and streamlined base fit right in with the movement to improve fan design. The fan is two speed and finished in a dark brown wrinkle finish. The fan will continuously oscillate when in operation.

Grade I – $100.00
Grade II – $75.00
Grade III – $50.00
Grade IV – $35.00
Grade V – $20.00
Grade VI – $5.00

Right: ***Roto-Beam***
Type: 1S3
Blades: 12"
Date: c.1938

Roto-Beam manufactured very futuristic fans for commercial and home use. This example is for home use and incorporates these futuristic designs. The base is made of stamped steel and finished in a gloss back. The guard is made of formed steel wire finished in polished chrome. The blades are made of cast aluminum and is forged from one piece. The motor is small and is attached to the guard. The motor seems to almost float by itself.

Grade I – $90.00
Grade II – $68.00
Grade III – $45.00
Grade IV – $32.00
Grade V – $18.00
Grade VI – $5.00

The Signal Electric Company of Menominee, Michigan, made many fans under their name as well as under the Cold Spot name for Sears. These fans were of modest quality and were designed primarily for the low priced fan market.

Left: ***Signal***
Blades: 10"
Date: c.1937

This fan has a stamped steel base and motor housing and is modestly constructed. The fan has a light green finish.

Grade I – $60.00
Grade II – $45.00
Grade III – $30.00
Grade IV – $21.00
Grade V – $12.00
Grade VI – $3.00

Center: ***Signal-Junior***
Blades: 10"
Date: c.1929

This fan has a universal motor and will oscillate. It has a light green finish.

Grade I – $60.00
Grade II – $45.00
Grade III – $30.00
Grade IV – $21.00
Grade V – $12.00
Grade VI – $3.00

Right: ***Signal***
Type: Cold Spot, Type 561,
Blades: 10"
Date: c.1938

The most outstanding feature of this fan is the guard which is deco in style. The fan has a silver finish making it very attractive.

Grade I – $50.00
Grade II – $38.00
Grade III – $25.00
Grade IV – $18.00
Grade V – $10.00
Grade VI – $3.00

Left:
Type: unknown
Blades: 8"
Date: c.1938

A fan with a slag-glass base. Glass is a very unusual material to use for a fan base.

Grade I – $90.00
Grade II – $68.00
Grade III – $45.00
Grade IV – $32.00
Grade V – $18.00
Grade VI – $5.00

Center:
Type: Cold Spot
Blades: 8"
Date: c.1929

A small oscillator finished in dark green. The blades are painted gold, alternating current only.

Grade I – $40.00
Grade II – $23.00
Grade III – $20.00
Grade IV – $14.00
Grade V – $8.00
Grade VI – $2.00

Right:
Type: Cold Spot
Blades: 6"
Date: c.1927

A small desk fan finished in dark green.

Grade I – $30.00
Grade II – $23.00
Grade III – $15.00
Grade IV – $10.00
Grade V – $6.00
Grade VI – $2.00

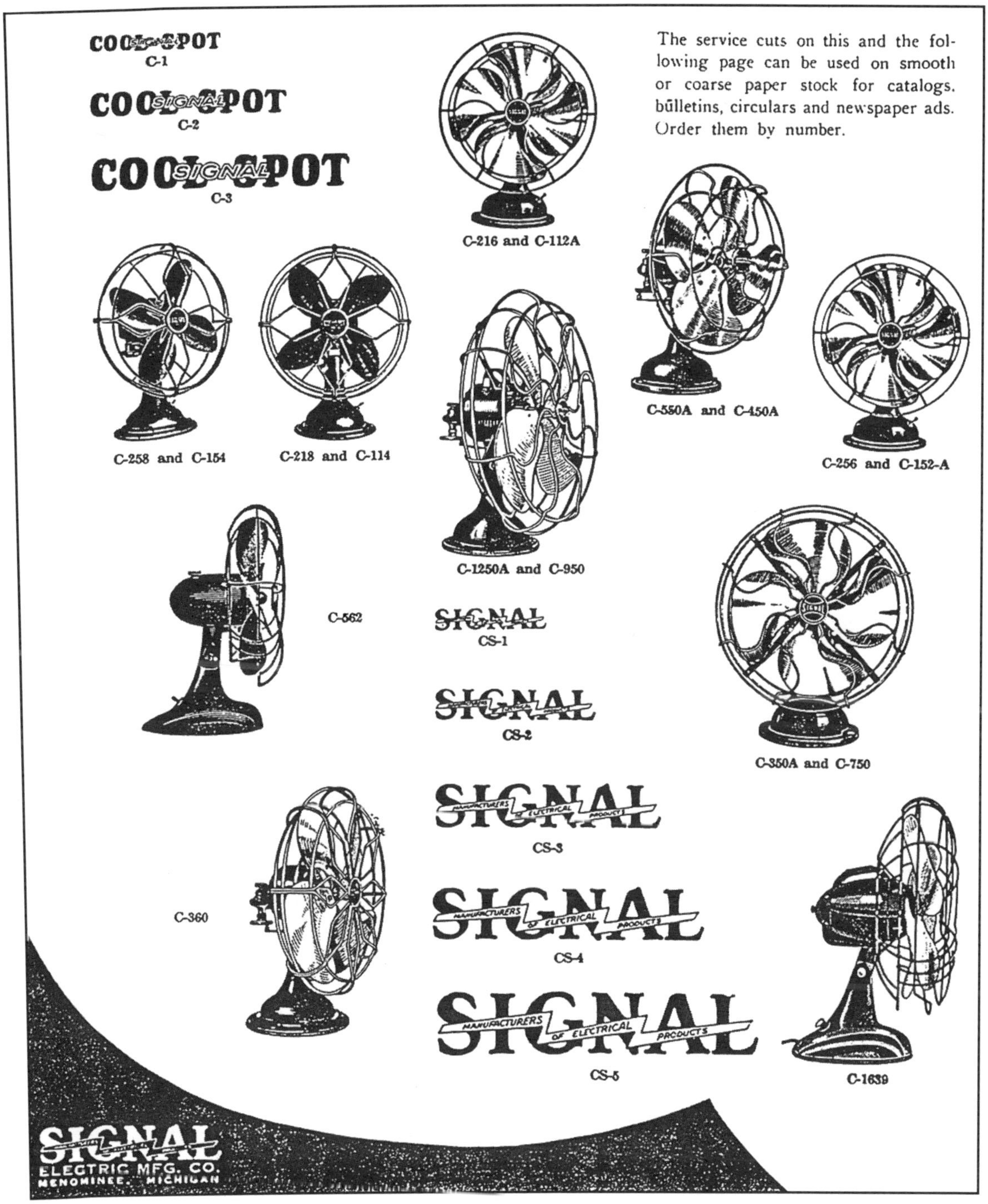

An illustration of Signal fans manufactured in 1940. Signal manufactured several grades of fans for home and commercial use. They also made fans for direct current.

SIGNAL

MANUFACTURERS OF ELECTRICAL PRODUCTS

10-INCH OSCILLATOR FAN

•

No. 550-A A.C. Only
***No. 450-A Uni. or D.C.**

List Prices

No. 550-A	$12.95
No. 450-A	13.50

*Operates on 110-120 Volts D.C. or A.C., 25 to 60 Cycles

Specifications

CODE WORD: No. 550-A—SEMAX.
No. 450-A—SEMAY.

MOTOR: No. 550-A, induction, 4-pole, non-radio interfering. Operates on 110-120 volts, alternating current, 60 cycles only. No. 450-A, universal, operates on 110-120 volts D. C. or A. C., 25 to 60 cycles.

BEARINGS: Self-lubricating with wool-packed oil reservoir, self-aligning.

BASE: Cast iron with felt base pad. Can be wall mounted.

OSCILLATOR: Adjustable for non-oscillating operation.

BLADES: Quiet type.

GUARD: Triple ring of heavy gauge wire, cadmium plated.

SWITCH: Two speeds, lever type in base.

CORD: Rubber, 8 feet.

AIR DELIVERY:
No. 550-A—700 C. F. M.
No. 450-A—740 C. F. M. on D. C., 570 on A. C.

WATTS: No. 550-A, 50. No. 450-A, 65.

FINISH: Black baked enamel with chrome plated blades and cadmium plated guard.

PACKING: 1 to a carton; 6 to a crate.

SHIPPING WEIGHT:
No. 550-A, Carton, 11 lbs. Crate, 83 lbs.
No. 450-A, Carton, 9¾ lbs. Crate, 73 lbs.

SPECIAL VOLTS AND CYCLES: No. 550-A. For voltages and/or cycles other than 110-120 volts, 60 cycles, add 10% to list price. Not made in frequencies less than 50 cycles. (See type 450-A for less than 50 cycles.)

No. 450-A—This standard fan (brush type) will operate on 25 to 60 cycles A. C. 110 volts or D. C. 110 volts. For voltages other than 110-120 add 10% to list price.

SIGNAL
ELECTRIC MFG. CO.
MENOMINEE, MICHIGAN

UL

An illustration of a top-of-the-line Signal fan from 1940. This fan had large overlapping chrome-plated blades that made the fan quiet to operate. This example is one of the more commonly found Signal fans.

Left: ***Signal***
Blades: 10″
Date: 1940

In 1940 Signal Electric came out with a line of desk fans with newly designed blades and guard. This new fan had highly polished chrome-plated blades and an improved guard. The motor, base, and oscillator remained unchanged from their fans of the late thirties.

Grade I – $60.00
Grade II – $45.00
Grade III – $30.00
Grade IV – $21.00
Grade V – $12.00
Grade VI – $3.00

Right: ***Signal-Junior***
Blades: 10″
Date: c.1939

This fan has a universal motor and is an oscillator. This has a light green finish.

Grade I – $50.00
Grade II – $38.00
Grade III – $25.00
Grade IV – $18.00
Grade V – $10.00
Grade VI – $3.00

Adams Bagnal
Type: Jandus No. 48091 C frame oscillator
Blades: 12"
Date: 1912

In 1912 the Adams Bagnal Company introduced their C frame oscillator. This fan used a track located in the frame that held the motor to oscillate the fan. The oscillator was very unusual design and worked quite well. The motor and base are made of cast iron finished in gloss black. The blades and guard are made of brass. A two-speed slide switch is located in the base.

Grade I – $4,000.00
Grade II – $3,000.00
Grade III – $2,000.00
Grade IV – $1,400.00
Grade V – $800.00
Grade VI – $200.00

Adams Bagnal
Type: Jandus No. 36095
Blades: 12"
Date: c.1912

Adams Bagnal also made non-oscillating desk fans for consumer use. This fan has an interesting wire trunnion which is also found on their gyro ceiling fans. The motor is made by General Electric. The guard and blades are made of polished brass. The center badge is made of cast brass and is very impressive.

Grade I – $900.00
Grade II – $675.00
Grade III – $450.00
Grade IV – $315.00
Grade V – $180.00
Grade VI – $45.00

Left:
Type: Kwik-Kool
Blades: 10"
Date: c.1928

This is a very inexpensive lower priced fan with a cast-iron base and stamped steel motor housing. The blades are made of stamped steel painted green to match the body of the fan.

Grade I – $30.00
Grade II – $23.00
Grade III – $15.00
Grade IV – $10.00
Grade V – $6.00
Grade VI – $2.00

Center:
Type: ZIP
Blades: 10"
Date: c.1930

This fan has a fancy ribbed cast-iron base and stamped steel motor housing. The blades and guard are painted silver. This fan will oscillate.

Grade I – $30.00
Grade II – $23.00
Grade III – $15.00
Grade IV – $10.00
Grade V – $6.00
Grade VI – $2.00

Right:
Type: ZIP
Blades: 10"
Date: c.1935

This fan has an unusual guard and badge. The blades are finished in a wrinkle finish which is also unusual for a fan. The fan will oscillate and is finished in lime green.

Grade I – $30.00
Grade II – $23.00
Grade III – $15.00
Grade IV – $10.00
Grade V – $6.00
Grade VI – $2.00

Victor

The Victor Company manufactured lower grade fans for commercial and home use. They seemed to be a mixture of several fan styles and the motors were often industrial in design and look. The hallmark of these fans are the breeze spreaders that were placed on the front of the guard deflecting the moving air. These breeze spreaders were also used on their ceiling fans making them very unusual.

Left: ***Victor***
Type: unknown
Blades: 16"
Date: c.1940

This large fan has overlapping blades that are similar to Emerson Parker and Signal blades. The motor is very crudely constructed and sits inside an oscillator ring like the General Electric fans of the twenties. There is a two-speed toggle switch in the base that is also similar to GE fans.

Grade I – $150.00
Grade II – $113.00
Grade III – $75.00
Grade IV – $53.00
Grade V – $30.00
Grade VI – $8.00

Right: ***Victor***
Type: unknown
Blades: 16"
Date: c.1936

Most Victor fans have six aluminum blades and this is an example. The guard is the S type held to the motor by steel struts. The motor sits in an oscillating ring on a stamped steel base. This example is missing its breeze spreader.

Grade I – $170.00*
Grade II – $128.00*
Grade III – $85.00*
Grade IV – $60.00*
Grade V – $34.00*
Grade VI – $ 9.00*

*This price would include the breeze speader.

Dayton

The Dayton Fan and Motor Company has been making fans since the turn of the century and still manufactures fans today. They produced high quality fans for both home and industrial use.

Left: ***Dayton***
Type: 22
Blades: 8"
Date: c.1910

An early Dayton fan with cast-iron motor housing and tab-foot base. The motor is universal and there is a three-speed slide switch in the base. The blades are made of sheet brass in the windmill style and the guard is brass with a cast-brass badge.

Grade I – $900.00
Grade II – $680.00
Grade III – $450.00
Grade IV – $320.00
Grade V – $180.00
Grade VI – $50.00

Left & Right : ***Dayton***
Type: 367
Blades: 10" and 12"
Date: c.1920

Both of these fans have a stamped steel base, cast-iron motor housing, and brass blades. The guard is made from formed steel in the S style. Many of these fans have a die-cast oscillator gear box that are joined between the motor and base. Unfortunately these parts tend to shatter if damaged and reduce the flexibility of the fan.

10" Fan
Grade I – $50.00
Grade II – $38.00
Grade III – $25.00
Grade IV – $18.00
Grade V – $10.00
Grade VI – $3.00

12" Fan
Grade I – $70.00
Grade II – $53.00
Grade III – $35.00
Grade IV – $25.00
Grade IV – $25.00
Grade VI – $3.00

Left: ***Crocker/Curtis***
Blades: 9"
Date: 1908

Right: ***Dayton***
Type: Type 21
Blades: 9"
Date: 1908

Crocker/Curtis made fans as far back as the late 1880s. This is an example of an all brass fan finished in polished brass. The universal motor is compact and high speed. There is a slide switch in the base. The guard is very unusual and has a cast-brass badge.

Grade I – $700.00
Grade II – $525.00
Grade III – $350.00
Grade IV – $245.00
Grade V – $140.00
Grade VI – $35.00

A very early fan with an open ring guard. The motor housing and base are made of heavy cast iron. The motor is designed for direct or alternating current.

Grade I – $700.00
Grade II – $525.00
Grade III – $350.00
Grade IV – $245.00
Grade V – $140.00
Grade VI – $35.00

Montgomery Ward, like Sears, Western Electric, and Singer never actually manufactured fans but purchased them from different fan manufactures and then put their name on them. This fan is such an example.

Left: ***Montgomery Ward***
Type: 86K5324
Blades: 12"
Date: c.1945

This fan has a cast-iron base, stamped steel motor housing, and is two speed. The blades and guard are chrome-plated steel in a modern style. This fan may have been manufactured by Knapp-Monarch or some other small fan company.

Grade I –	$70.00
Grade II –	$53.00
Grade III –	$35.00
Grade IV –	$25.00
Grade V –	$14.00
Grade VI –	$3.00

Right:
Type: Kwik-Kool
Blades: 9"
Date: c.1945

Small oscillating fan with chrome-plated guard and blades. This is a plug in fan with a cast-iron base. The paint is a wrinkle finish.

Grade I –	$30.00
Grade II –	$23.00
Grade III –	$15.00
Grade IV –	$10.00
Grade V –	$6.00
Grade VI –	$2.00

Left:
Type: Carleton "round ball"
Blades: 8"
Date: c.1910

This fan has a nickel-plated guard and blades with a round ball iron motor. The base is cast iron. A very attractive fan.

Grade I –	$500.00
Grade II –	$375.00
Grade III –	$250.00
Grade IV –	$175.00
Grade V –	$100.00
Grade VI –	$35.00

Center:
Type: Pittsburgh, Type 6
Blades: 6"
Date: c.1922

A very small fan with a stamped steel motor housing and base. The blades are brass in the windmill style.

Grade I –	$300.00
Grade II –	$225.00
Grade III –	$150.00
Grade IV –	$105.00
Grade V –	$60.00
Grade VI –	$15.00

Right:
Type: Busy "B"
Blades: 6"
Date: c.1922

A small non-oscillating fan.

Grade I –	$30.00
Grade II –	$23.00
Grade III –	$15.00
Grade IV –	$10.00
Grade V –	$6.00
Grade VI –	$2.00

Left:
Type: Aeriet Fan
Blades: 14"
Date: c.1947

Right: ***Kenmore***
Type: Model 124.8055
Blades: 12"

This is a very unusual fan in the post-World War II modern style. The motor is attached to the base by a steel wire guard. The motor is rather large and has a small cooling fan located below the stator. This small interior cooling fan is an unusual feature for a fan motor. The base and motor housing are finished in a dark green wrinkle color. The blades are shaped in an unusual configuration and are attached to the motor by a large flat chrome button. There is no switch in the base or on the motor, an unusual feature for such a large fan. This fan is very stylish by not very safe with such an open guard.

Grade I – $90.00
Grade II – $68.00
Grade III – $45.00
Grade IV – $32.00
Grade V – $18.00
Grade VI – $5.00

This is a very modern and stylish post-World War II fan incorporating new post war technological improvements. It has a high speed motor atached to high velocity blades. It is very similar to Vornado fans of the period and incorporates similar aerodynamic technology. The base and large ring around the motor are dark metallic brown, the blades and fan housing are light metallic brown. The blades are deeply pitched and squared off. There is a two-speed toggle switch located at the front of the base. This is a very powerful efficient fan marketed by Sears.

Grade I – $70.00
Grade II – $53.00
Grade III – $35.00
Grade IV – $25.00
Grade V – $14.00
Grade VI – $3.00

Left: ***Surf***
Type: Cat. #540-1
Blades: 10"
Date: c.1952

Right: ***Vornado***
Type: unknown
Blades 10"
Date: c.1955

Following World War II many small fan companies emeraged marketing fans for consumer consuption. The Surf fan is one of the most interesting. This fan incorpoarates post-war modern design having simple and uniform lines. The base is made of a simple peice of formed steel with a brown wrinkle finish. The motor is attched to a wire guard which is attched to a simple trunnion. The motor is a simple record player motor surrounded by a stamped metal shroud. The blades are overlapping and have a shallow pitch. This fan fits in to the simple lines assocated with the modern period following World War II.

Grade I – $30.00
Grade II – $23.00
Grade III – $15.00
Grade IV – $10.00
Grade V – $6.00
Grade VI – $2.00

Vornado manufactured a new concept fan using aerodynamic technology used in airplane design. The Vornado company touted that their product was more efficient than the old fashion traditional buzz fans and had style to boot. These fans were very expensive when new, and mixed both industrial and residential design. This fan has a light blue finish with black plastic blades. The base is made of stamped steel with a two-speed rotary switch. The blades rotate at a high velocity making the fan rather noisy. Vornado still manufactures high quality fans today.

Grade I – $70.00
Grade II – $53.00
Grade III – $35.00
Grade IV – $25.00
Grade V – $14.00
Grade VI – $3.00

Left: ***Air Flow***
Type: Cat #2426
Blades: 10"

Right: ***Sterling***
Type: ASU
Blades: 9"

A Depression-era fan with six stamped steel blades. It is unusual for a 10" fan to have six blades. The base and motor housing are made of stamped steel and there is a small handle on top of the motor for easy transport. This fan will oscillate.

Grade I – $30.00
Grade II – $23.00
Grade III – $15.00
Grade IV – $10.00
Grade V – $6.00
Grade VI – $2.00

A Depression-era fan with a cast-iron ribbed base. The base and motor housing are finished in high gloss green with a large black metal strip holding the motor to the base. The blades are made of stamped steel and are surrounded by an S wire guard.

Grade I – $20.00
Grade II – $15.00
Grade III – $10.00
Grade IV – $7.00
Grade V – $4.00
Grade VI – $1.00

Glossary

AC – alternating current

badge – an identifying name plate found on the front of most guards and is usually round in shape

Bakelite – an early plastic used in the middle to late thirties

bullet back – a cover placed over the back of a fan motor to give a streamlined appearance

continuous oscillator – an oscillator that does not have an on/off device and makes the fan go back and forth when in operation

DC – direct current

Deco – a shortened term for Art Deco, a style of art dating from the late twenties and thirties

die cast – a process of processing metal used in the teens and twenties to manufacture fan parts

guard – wires around the fan blades to protect the blades from damage

modern style – a period of art and design staring around World War II and continuing into the early fifties

overlapping blades – large blades that tend to have a deep pitch and cover a large area, found on fans after the middle thirties

ornate base – most commonly used to describe Emerson fan bases that are stepped and ribbed

oscillator – a mechanism that makes a fan go back and forth sending out air in different directions

pancake motor – a flat, thin motor used to describe fan motors found on some early fans, especially General Electric before 1907

Parker blade – a fan blade designed by H. Parker, president of Emerson Electric, and patented in 1899

ribbed base – a cast-iron base with small vertical ribbing used on early fans

rotary switch – an on/off mechanism used by turning the control knob in one direction, placed most often on the base

S type guard – a guard with wired forming an S shape, a standard feature on most early fans

sidewinder oscillator – an oscillator that has levers extending from the side of the gearbox to the base that push and pull the fan back and forth

slide switch – an on/off mechanism used on most fans before the thirties

star oscillator – an oscillating mechanism manufactured by General Electric around World War I that has brass set screws with faceted tops

tank motor – a large heavy cast-iron motor used on early fans

toggle switch – an on/off mechanism that will go back and forth from one point, used on later fans

trunnion – a bracket designed to hold a motor to the base and allow the movement up and down

universal motor – an electric motor that can uses either AC or DC power and will have brushes

vane oscillator – a device that uses air to push the fan back and forth, an early form of oscillation

windmill blades – fan blades that are squared off at the edge forming a triangle shape, used on early fans before World War I

Index